SOIL ANIMALS

Ann Arbor Science Library

Soil Animals

by Friedrich Schaller

ANN ARBOR
THE UNIVERSITY OF MICHIGAN PRESS

Library of Congress Catalog Card No. 68-13663
Published in the United States of America by
The University of Michigan Press and simultaneously
in Rexdale, Canada, by Ambassador Books Limited
United Kingdom agent: The Cresset Press
First published as Die Unterwelt des Tierreiches
(Verständliche Wissenschaft, Band 78) © by
Springer-Verlag OHG Berlin-Göttingen-Heidelberg 1962
Translated by A. J. Pomerans
Manufactured in the United States of America

PREFACE

It is characteristic of biology that its problems are capable of treatment on several levels and by several distinct methods. Its subject matter ranges from description and classification of organisms to the analysis of their history and their life processes. Although the first two disciplines—morphology and taxonomy—are considerably older than the last two—genetics and physiology—all of them continue to engage the attention of modern biologists.

Zoologists alone have to deal with more than a million known species, and there is every reason to suppose that several hundred thousand more will be added as developing countries are opened up to zoological research. Even in our own latitudes the description and classification of species is far from completed, particularly when it comes to the smallest species.

Until only a few years ago, the life and behavior of the tiny inhabitants of the soil were among the least-known chapters of "descriptive" biology. True, it was generally appreciated that the world beneath our feet was teeming with worms, insect larvae, primitive insects, millipedes, mites, protozoa, and the like, but the life of these light-shy organisms remained an almost closed book.

Before it could be opened, new methods of catching, breeding, and observing the soil animals had to be developed, and these techniques, together with the secrets they have revealed, will be among the subjects discussed in this book. Since the average reader is familiar with no more than a few of these animals, we shall begin our account with a brief description of the most important among them. Then we shall look at their distribution and biological importance. Chapters III and IV will deal with their general habits and life. In so short a volume, I have been able to give only a few examples, but these suffice to show that life in the dark and narrow confines of the soil is by no means as uniform or monotonous as one might expect. Indeed, quite the contrary is true, as the chapter on the sexual biology of soil animals will try to make clear.

I am indebted to my pupils for their indispensable help in the preparation of the chapters dealing with the habits of soil animals and for the many photographs and diagrams explaining the strange mating rituals of soil animals. Readers interested in further taxonomic and soil-biological details are referred to the *Soil Biology* of my teacher, Wilhelm Kühnelt (London, 1961).

Friedrich Schaller

CONTENTS

I
INTRODUCTION

In its natural state the soil is not only a mixture of decayed mineral and organic substances, but is also the home of many animals and plants. Taller plants divide the soil with their roots; their cavities are inhabited by a host of animals and lined with fungi, algae, and bacteria. Many people are familiar with some of the larger soil-dwelling creatures: moles, voles, various beetle larvae, mole crickets, and earthworms being among the commonest. They all bear clear signs of their ancestry; they are either worm-shaped, which helps them to force their way through the soft ground, or are specially equipped for burrowing, as for instance the mole and the mole cricket (Fig. 1b). Far richer in species and individuals are those groups that cannot actively burrow through the soil and must depend on the existing channels. Most people know very little indeed about the life of these creatures, and even soil experts and biologists were rather unfamiliar with them until a few decades ago. What they have learned since, however, is intriguing enough to excite even the nonprofessional reader, because of the profusion of forms and modes of life that have been discovered, because of the biosynthetic importance of the organisms involved, and because of their unexpectedly varied and curious behavior patterns.

Collecting Small Soil Dwellers

Most soil-dwelling animals are so small that they are easily missed. Perhaps that is just as well, for which of us, walking in a wood or meadow, would wish to know that every step endangered hundreds of tiny lives? Even when digging our garden, we turn up no more than the

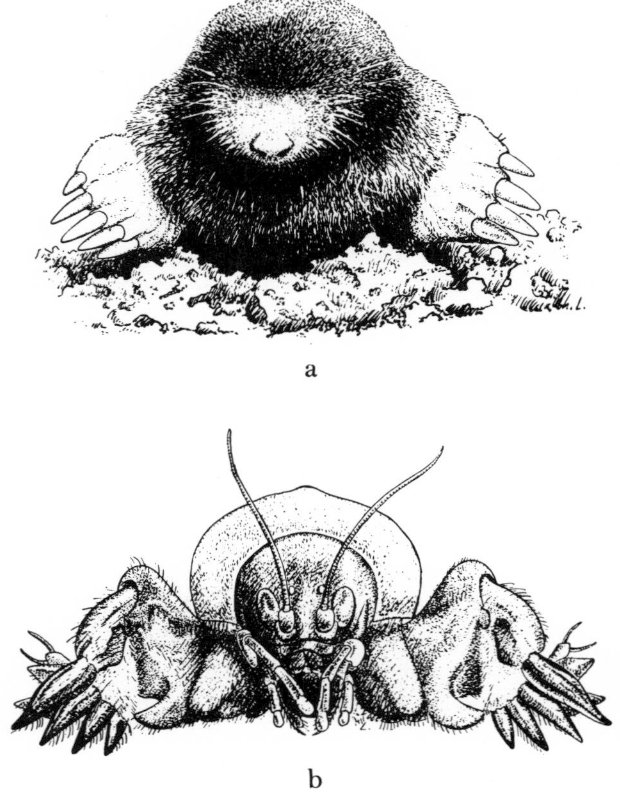

FIG. 1. (a) Mole. (b) Mole cricket. Note the identical development of the forelegs as burrowing organs and the similarity in shape of two completely unrelated species ("convergence," see p. 75).

occasional grub or an earthworm; apart from these, the soil seems quite devoid of animal life. Of course, things look different when we view a handful of fresh humus through a powerful lens. Usually, the teeming animals in it quickly disappear below the surface, and our little sample looks deserted once again, for there is nothing these creatures shun more than light and dryness. In that respect they resemble troglodytes; many of them are colorless and blind. We observe them groping their way forward with their legs and feelers and slipping into the narrowest of gaps. The more robust among them, however, may be caught on the surface with a suction pipe (Fig. 2).

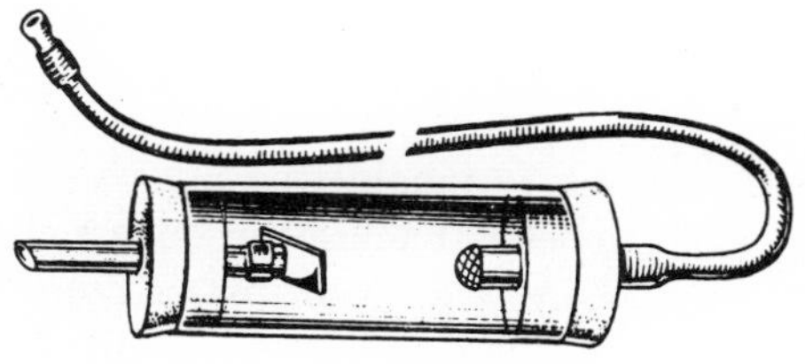

FIG. 2. An aspirator, or suction pipe, for the collection of small soil animals (from Balogh, 1958).

To get an idea of the diversity of these creatures, we must resort to a dodge, first introduced by Berlese, a famous nineteenth-century Italian zoologist. Knowing that soil-dwelling animals abhor light, dryness, and heat, he used these properties to drive them from their underworld (Fig. 3). A sample obtained by this method will show us, too, how packed with living creatures the soil really is. Yet this is but a fraction of the actual wealth of individuals and types—the Berlese funnel picks up only the more mobile forms, especially insects and their larvae, arachnids, and myriopods. Other animals, moving more laboriously, for instance nematode worms or crea-

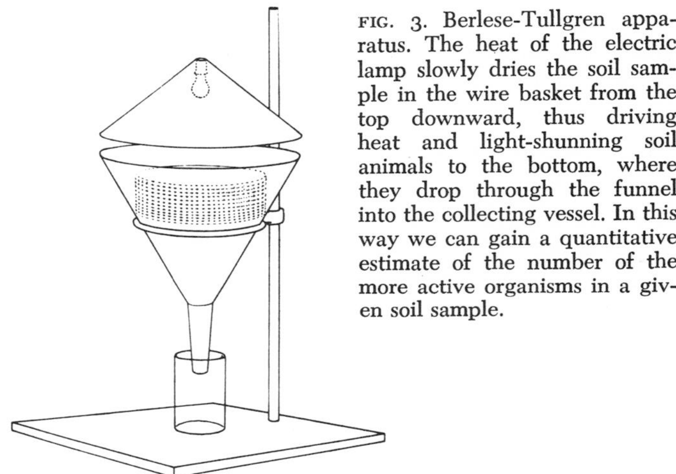

FIG. 3. Berlese-Tullgren apparatus. The heat of the electric lamp slowly dries the soil sample in the wire basket from the top downward, thus driving heat and light-shunning soil animals to the bottom, where they drop through the funnel into the collecting vessel. In this way we can gain a quantitative estimate of the number of the more active organisms in a given soil sample.

tures small enough to inhabit the invisible film of water covering soil particles, enclose or encyst themselves in self-secreted capsules and await better times on the spot. In general, this solution proves successful, for in nature—unlike the artificial Berlese environment—aridity eventually gives way to rain. In any case, different methods must be used to capture these ingenious creatures; the most reliable technique is to suspend tiny particles of soil in water and to "comb" them microscopically. Unfortunately, this is an extremely time-consuming process, and biologists often prefer to isolate species such as the nematode worms by means of special extractors (Fig. 4).

It is not possible to make a full and accurate count of all the species and individuals contained in a given sample of soil; all we can do is to make comparative counts of the three most important classes:

a) large and robust forms, more than 1 cm* in length, which can be sifted from coarse samples (earthworms, larger insects and their larvae, myriopods, and arachnids);

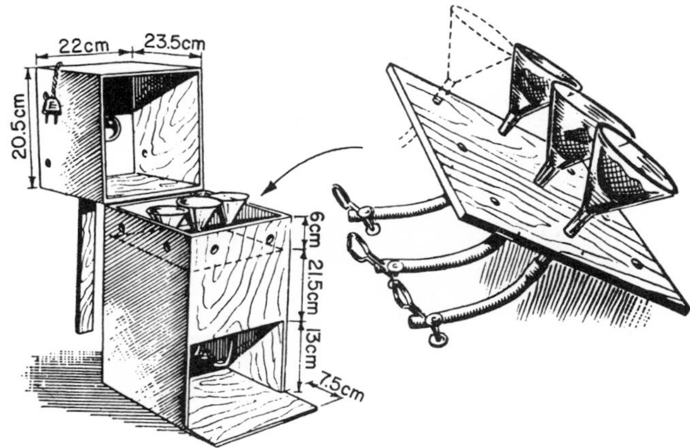

FIG. 4. Apparatus for collecting nematode worms. Small samples of soil (1cc-4cc) are placed on trays of wire gauze inside funnels filled with water at 13° C. A 16-watt lamp is used to heat the water to 30° C. As the temperature rises, the worms begin to emigrate from the soil samples. At 30° C they lapse into a state of torpor and drop down the funnel until, after some 12 hours, they can be collected at the bottom of the funnel tubes (from Overgard-Nielsen, in Balogh, 1958).

b) small, mobile forms, 0.5-5 mm in length, which are best collected by means of the Berlese funnel;

c) microscopic moist-soil organisms, which must be examined under the microscope in moist samples of soil.

Brief Survey of Soil Animals

From what soil zoologists have discovered by the methods described, we know that, with the exception of sponges, coelenterates, mollusks, echinoderms, fishes, and birds, there is no large group of animals that has not evolved soil-dwelling species of its own.

*1 mm = 0.04 inches (a twenty-fifth of an inch); 1 cm = 0.39 inches (approximately two-fifths of an inch); 1 dm = 3.94 inches; 1 m = 39.37 inches.

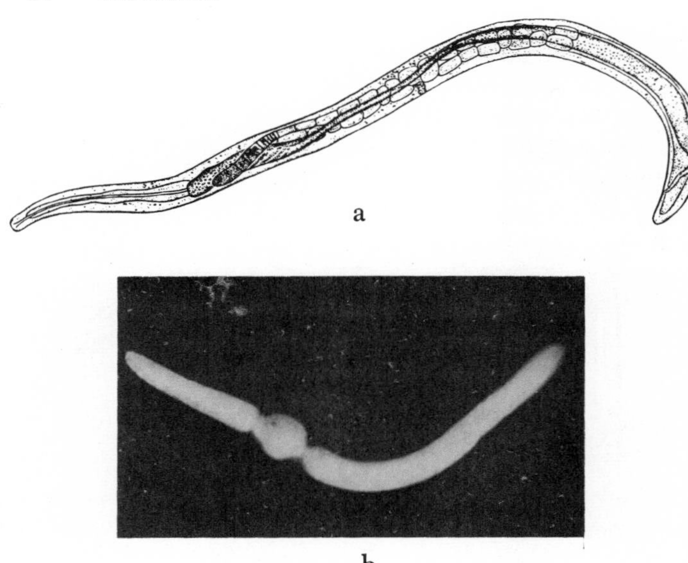

FIG. 5. Soil worms. (a) Roundworm, nematode (*Dorylaimus papillatus*). Actual length about 1 mm (from Kükenthal, 1933). (b) Enchytraeid worm, a smaller relative of the earthworm. Actual length about 1 cm (from Trappmann, 1954).

Worm-shaped soil inhabitants (Fig. 5) either propel themselves forward by peristalsis—by contracting their bodies in successive circles while bracing themselves against the soil with their bristles—or by eating their way through the soil, as earthworms do, or by wriggling through the moist substrate, as do roundworms. Next to the microscopic protozoa, roundworms (nematodes) are the most widely distributed creatures in the soil—there are 1000 to 10,000 of them to every cubic centimeter. As a rule, they secrete a slimy substance that helps their progress. More often than not they settle among roots in the upper layers of the soil. If this layer dries out, they readily enclose themselves in a cyst or capsule, in which form they can survive for long periods. Then, as soon as the organic matter in their immediate environment begins to decompose, they reappear in enormous numbers.

Nematode worms share their ability to form cysts with their relatives, the rotifers and tardigrades. These minute soil and moss dwellers are incredibly tough and, enclosed as cysts, can even withstand long immersion in liquid helium (—271° C.).

Soil-dwelling snails are characterized by the exceptional frailty of their shells.

Arthropods include the greatest number of species among soil creatures. Even the lower crustaceans, normally associated with an aquatic form of life, boast numerous true soil dwellers. These members of the subclass Copepoda (*Cyclops, Calanus*) normally live among wet foliage (Fig. 6). The characteristically long antennae of most copepods are very short in the soil forms, which do not skip through water but must wriggle through the soil on their bellies. Like other crustaceans, they provide particularly convincing evidence that many soil inhabitants are descended from aquatic ancestors. This is most obvious in the wood lice, found in all types of soil with

FIG. 6. Minute soil-inhabiting crustacean (copepod) enclosed in a cyst. Natural length 1 mm (from Kühnelt, 1950).

suitable pores. Instead of the original gills, wood lice have lunglike infoldings (invaginations) on their abdominal appendages (hind legs); when filled with air these appear as prominent white spots. The wood louse shape is very common among medium-sized soil animals, and the little creature's ability to curl or roll up—which has

FIG. 7. Wood louse (isopod). Actual length 1.5 cm.

earned one of its relatives the name of pill bug—is found among other soil inhabitants as well. Many wood lice do not so much live *in* as *on* the soil, beneath stones, wood, and leaf litter (Fig. 7). In the next chapter we shall see that these "surface dwellers," too, are true soil inhabitants and that they play a very special part in the soil's life cycle.

Soil-dwelling animals include a host of arachnids, among them many groups not found in Central Europe, such as scorpions and whip scorpions (Pedipalpi). Together with pseudoscorpions, harvestmen or daddy longlegs, and soil spiders, they form the bulk of the predators in the soil community (Fig. 8).

Mites, too, are predominantly soil dwelling (Fig. 9), and there is no sample of humus that does not contain a mass of herbivorous beetle mites (oribatids). The same is true of springtails (Collembola)—an order of primitive insects—which, with the beetle mites, are among the most important producers of humus (Fig. 10).

The great majority of insects can be classified as soil animals, inasmuch as very many of them spend their

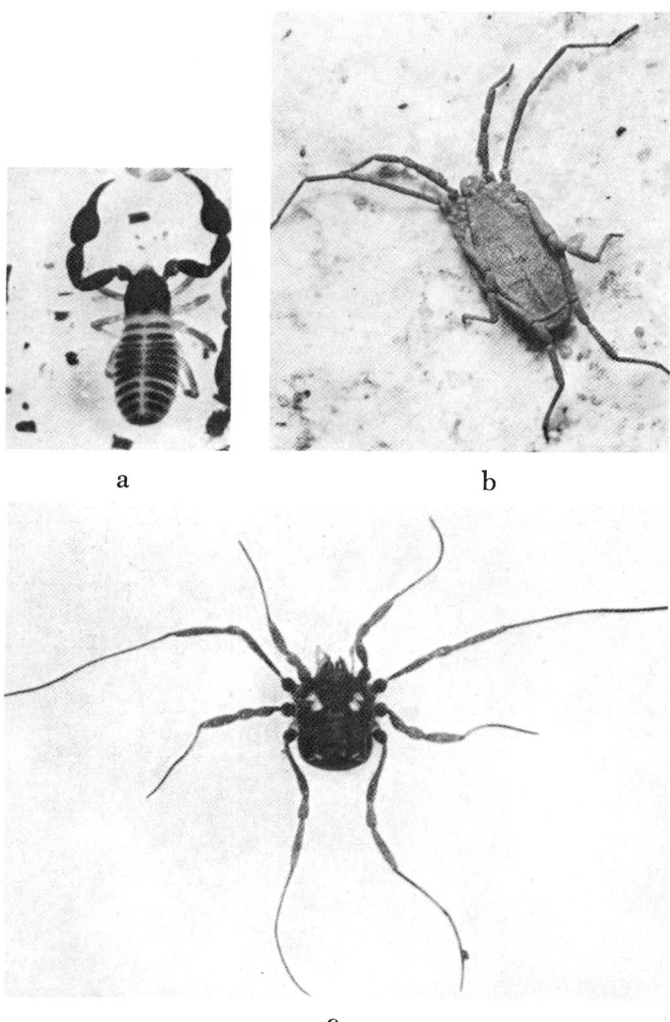

FIG. 8. Arachnids. (a) Pseudoscorpion. (b) Harvestman (*Trogolus*). (c) Spider (*Nemastoma*).

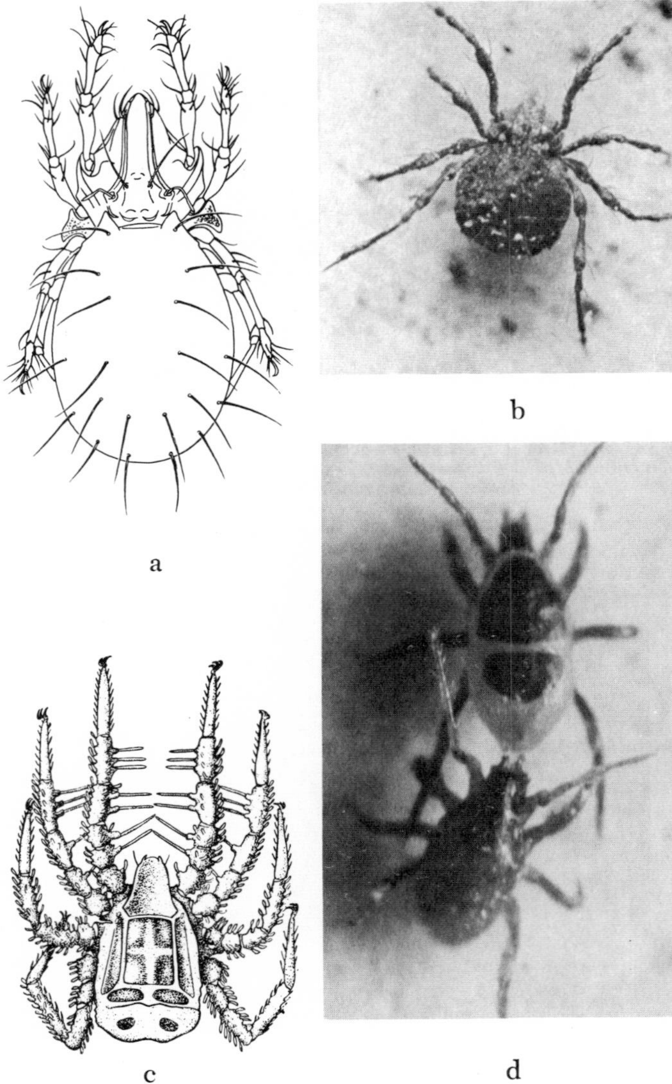

FIG. 9. Various soil beetle mites (oribatids). (a) *Tetracondyla.* (b) *Belba.* (c) *Caeculus echinipes,* a particularly striking oribatid (from Berlese, in Kühnelt, 1950). (d) Male and female of *Parasitus coleoptratorum* (from Rapp). The actual length of the mites is about 1-2 mm.

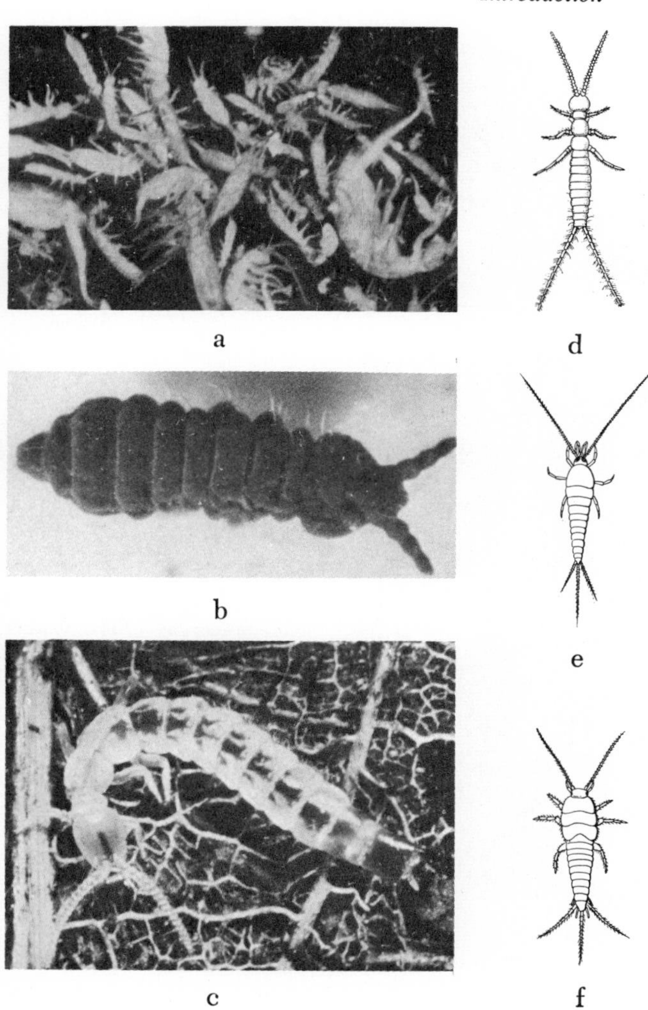

FIG. 10a-f. Springtails and other primitive insects. (a) Various species of springtails inhabiting upper layers of the soil. (b) Springtail (*Hypogastrura*) from middle soil layer. (c) *Japyx*, a predatory primitive insect from lower soil layer (actual size about 1 cm). (d) *Campodea* found in all soils rich in humus (actual size 1 cm). (e) Silverfish (*Machilis*) and (f) silverfish (*Lepisma*), both belong to the bristletails (order Thysanura).

(from A. Brauns, 1954).

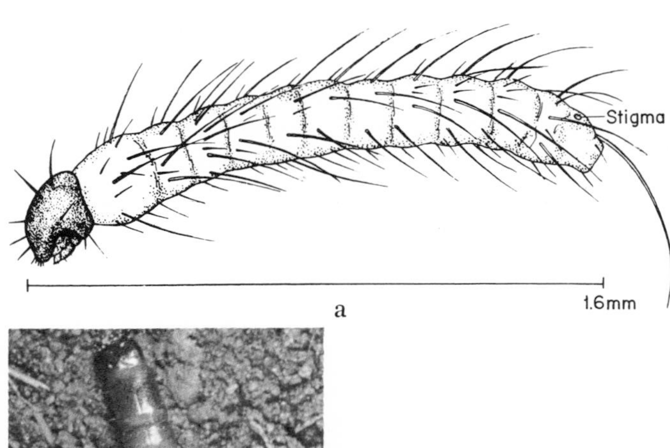

a

b

FIG. 11. (a) Young larva of St. Mark's fly (*Bibio marci*) (from A. Brauns, 1954). (b) Wireworm, the larva of the click beetle.

larval stage in the soil (Fig. 11). This applies to most beetles and cicadas, to a host of flies, midges, and bugs, and to numerous butterflies and moths. In addition, a considerable number of adult insects (imagos) live in the soil, among them many weevils and mole crickets.

Myriopods are soil creatures without exception; only a few of them leave the soil temporarily to feed on the algae of tree trunks or on other prey (Fig. 12).

Vertebrates, finally, are only a very few of the true

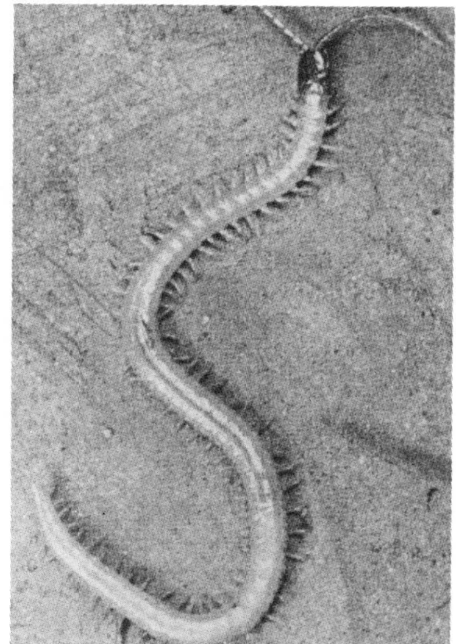

a

b

FIG. 12. Soil-inhabiting myriopods. (a) Carnivorous centipede (*Geophilus*). (b) A julid (vegetarian millipede).

soil dwellers. All of them are large and active burrowers, including the moles, voles, caecilians, shrews, and skinks.

Unicellular soil animals (Protozoa) deserve special consideration. They live exclusively in the film of moisture with which all hollows in the soil are lined. To that extent soil protozoa may be said to have remained aquatic animals. Some forms, however, live chiefly in the ground, and for that reason they may be classified as "true" soil animals.

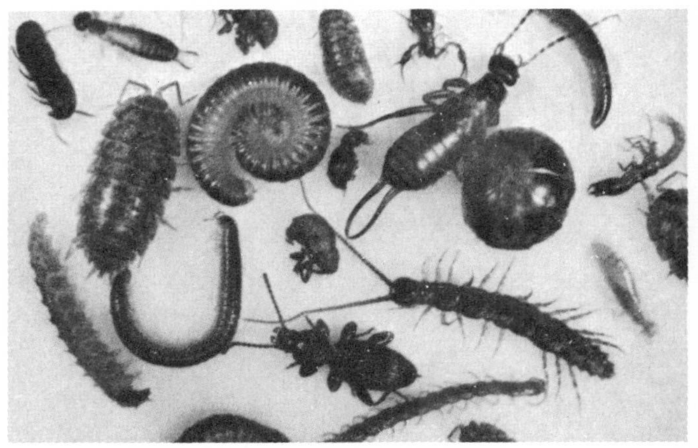

a

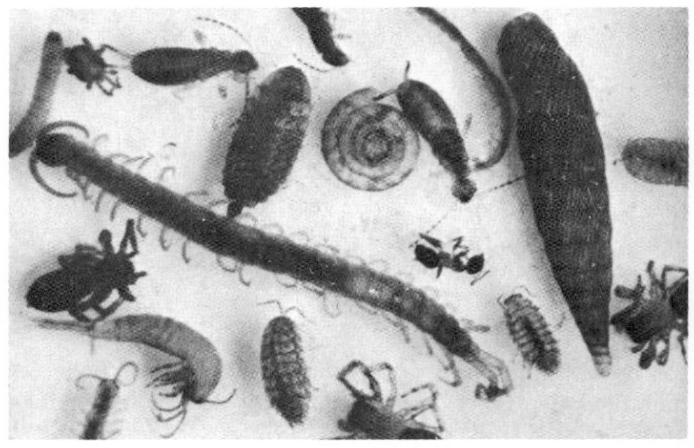

b

FIG. 13. Typical humus dwellers, excluding small species. (a) In beech-leaf litter in the Vienna Forest. (b) In the oak and hornbeam leaf litter in the same forest (from Kühnelt, 1950).

CLASSIFICATION OF THE MOST IMPORTANT SOIL ANIMALS (cf. Fig. 13)

1. Unicellular organisms (Protozoa):	Amoeba, naked and shell-bearing varieties Flagellates Ciliates
2. Worms:	Flat worms (Turbellaria) Rotifers Roundworms (Nematodes) Earthworms and enchytraeid worms (oligochaete annelids)
3. Tardigrades (Bear animalcules)	
4. Onychophora (*Peripatus*)	
5. Mollusks:	Snails and slugs
6. Arthropods:	Crustaceans: Terrestial copepods Isopods (wood lice) Arachnids: Scorpions Pseudoscorpions Harvestmen Soil spiders Mites Myriopods: Pauropods Symphyla Diplopods (millipedes): Julids (snake millipedes) Glomerids (pill millipedes) Pselaphognatha (tufted millipedes) Chilopods (centipedes): Geophilids Lithobids Scolopenders

	Insects:	Primitive insects:
		Protura
		Springtails
		Japygids
		Campodeids
		Machilids
		Silverfish
		Winged insects (Pterygota):
		The larvae of many insects and various adult forms such as earwigs, crickets, beetles, cockroaches, etc.
7. Vertebrates:	Amphibians:	Gymnophiona (limbless burrowing caecilians)
	Reptiles:	Amphisbaeridae (wormlike reptiles)
		Skinks
		Typhlopidae (snakes)
	Mammals:	Moles
		Pocket gophers
		Shrews
		Voles
		Armadillos

Characteristics of Soil Animals

A systematic classification of soil animals reveals that the various groups have little in common, so little in fact that we begin to wonder what if anything is "typical" of "true" creatures of the soil. The answer lies in the ecological field, that is, in a classification based on the relationship between animals and their environment rather than on structural (morphological) considerations. Ecological studies lead to the following divisions:

1) Large, actively burrowing animals that can move through the soil at will.

2) Extremely varied, medium-sized, surface and humus dwellers and those living under loose stones.

3) Small inhabitants of the loose upper and middle layers of the soil ("hemiedaphon"*).

4) Small and very small forms inhabiting deeper layers ("euedaphon"†).

5) Microscopic organisms inhabiting invisible films of moisture and damp substrata of the soil.

In the strictest sense, only members of Group 4 ought to be called "true" soil inhabitants. All of them are blind and lacking in pigment. In the dark, eyes would in any case be useless, and skin pigmentation would serve them neither as protection against the rays of the sun, nor as a means of attracting or identifying their equally blind neighbors. They have a thin covering of hair, short appendages (legs), and, as a rule, highly developed organs of touch and smell, and they are rarely more than 2 mm long. The smaller forms are mostly spherical; the larger ones worm-like in appearance and movement.

Members of Group 3, classified as hemiedaphic organisms, have rudimentary eyes; most of these creatures are slightly pigmented; many are heavily armored and capable of rolling up or curling into a ball, and of jumping.

Ecological diversity is clearly exemplified by springtails (order Collembola), which inhabit various layers of the soil and show the appropriate adaptations (Fig. 14).

Another system of classification is based on nutritional type. Herbivorous and carnivorous soil animals are readily distinguished by both appearance and behavior. The herbivores move slowly and easily; they are generally rounded and have rudimentary sense organs. The predators, on the other hand, are agile, have well-developed sense organs, and are specially equipped for hunting. The contrast is particularly clear when we compare diplopods with chilopods among the myriopods (cf. Fig. 10), or springtails (Collembola) with japygids (Diplura) among primitive insects.

*From Greek: *hemi* = half; *edaphos* = soil.

†From Greek: *eu* = good, true; *edaphos* = soil.

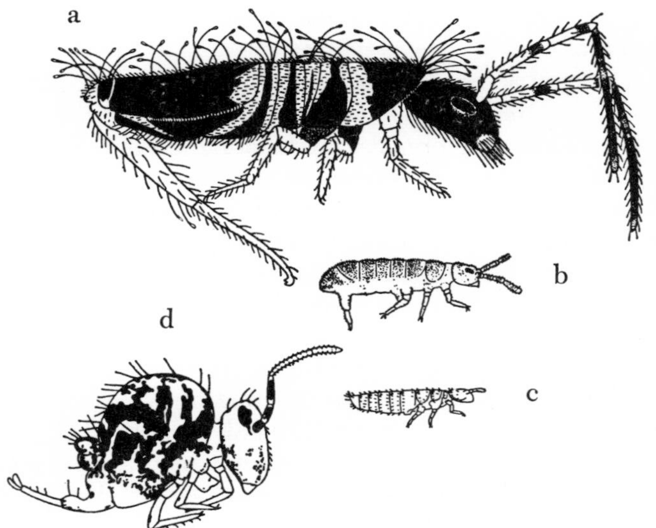

FIG. 14. The external appearance of springtails. (a) Surface inhabitant (*Entomobrya*); (b) inhabitant of topsoil (*Proisotoma*); (c) inhabitant of lower strata (*Tullbergia*); (d) a grass dweller (*Sminthura*).

We can also introduce the following subdivisions: animals feeding on 1) living plants, 2) dead vegetable matter, 3) fungi, 4) algae, 5) bacteria, 6) other animals, 7) dung, 8) carrion, and 9) detritus; parasitic animals make a tenth group. When it comes to the higher insects, however, this system of classification must be qualified, since, in the larval stage, insects often have quite different feeding habits than they have in the adult phase—we need only think of the owlet moth which feeds on flowers, while in its caterpillar stage it feeds on the roots of grasses and vegetables.

To appreciate just how diverse feeding types may be among a single group of soil dwellers, remarkably uniform in other respects, we have but to look at the nematode worms. Normally, these creatures have a narrow mouth with which they can suck up bacteria and organic fragments from damp substrata. Many of them are also semiparasitic—the mouth is armed with a hollow spear that enables the worm to penetrate the toughest root without seriously damaging the "host." Others have become notorious plant pests. Still others have tiny teeth or grinding plates with which they can pulverize bacteria. Finally, some nematodes are true predators and have large teeth with which they can attack and devour small worms, rotifers, and tardigrades (Fig. 15).

FIG. 15. A predatory nematode (*Diploscapter*) (from W. Kühnelt, 1950).

II

THE BIOSYNTHETIC IMPORTANCE OF SOIL ANIMALS

Ever since biologists learned just how densely populated the soil really is, they have been asking themselves what precise role the soil animals play in the development and metabolism of the soil. Traces of animal activity in the soil can be detected in: the shape of burrows, passages, and channels, the breakdown of vegetable matter, the form of excrement, refuse, and cadavers. In addition, biologists also consider an important "side" effect: the transport of materials.

Darwin and Earthworms

All this is best illustrated by the "classic" example of earthworms. Every schoolboy knows that earthworms are an extremely important link in the life cycle of nature, but not everyone realizes that this part of our biological education is largely based on the work of Charles Darwin. He was the first to show that earthworms loosen and aerate the soil by their ceaseless plowing activity, and that they enrich it by introducing plant segments (mainly leaves), by eating and digesting these together with mineral matter, and by depositing the resulting excrement (clayey humus) in their burrows and on the surface of the soil. The great density of earthworms in different soils was well known to Darwin, who even made quantitative estimates of them. In 1881 he counted some 23,000 earthworms per acre in English pastures; more recent investigations (Guild, 1951) have shown that the density in Scottish pastures is from 100,000 to 190,000 individuals per acre.

Kollmannsperger (1934) has given the following figures for various soils (after Kühnelt, 1950):

Type of land	*No. per square meter*	*Weight per square meter (in gm.)*
Gardens	approx. 390	72
Meadows	8-293	20-80
Fields	70-112	19-53
Woodland	44-74	16-28
Wasteland	7-11	0.3-0.4

Darwin also observed that earthworms covered a layer of red sand evenly distributed over the surface with 2 inches (5 cm) of earth in the course of 7 years, and that a layer of marl had "sunk" some 10-11 inches in the course of 28 years.

More precise measurements have been made, and the results have been summarized by Kühnelt (see Table 1). These surprising achievements by earthworms are partly due to their attaining a relatively old age (up to 10 years).

Darwin, who busied himself with the study of earthworms until his death, often showed visitors an impressive experiment designed to demonstrate the worm's sensitivity to vibrations. On his piano stood a whole row of flower pots, each containing an earthworm. At night they would come out to fetch leaves. Whenever he struck certain low notes, they would suddenly withdraw into their burrows; other notes would not affect them in the least.

Though Darwin was well ahead of his contemporaries even in the field of soil zoology, he, too, failed to appreciate how rich in soil animals the earth really is, so much so that it took his successors a long time to restore

TABLE 1. Excrement produced by earthworms in the course of 1 year under various climatic conditions

	Earthworm excrement in g/m² of soil near Zurich (clayey soil) (Stöckli)							Bellinchen (much drier climate) Heath (unshaded)
Month	*Garden*	*Meadow**	*Forest meadow*	*Orchard*	*Golf course (always moist)*	*Mixed forest*	*Pine grove*	
Jan.	. . .	. . .	. . .	. . .	. . .	. . .	. . .	. . .
Feb.	. . .	. . .	. . .	. . .	. . .	. . .	. . .	. . .
Mar.	6	135	200	180	140	49	20	4
Apr.	10	486	1370	530	1000	545	170	46
May	. . .	937	1400	390	725	469	460	36
June	. . .	917	1000	380	1235	309	194	17
July	. . .	244	260	305	689	115	63	35
Aug.	171	163	400	460	702	130	120	22
Sept.	350	403	1265	580	1364	247	283	210
Oct.	250	557	1030	260	1320	66	355	170
Nov.	100	486	384	115	643	50	328	20
Dec.	17	81	28	70	307	40	181	. . .
Total	904	4409	7837	3270	8125	2020	2174	560

* Annually treated with liquid manure.

zoology to its rightful place in soil studies. This process began with the discovery that the soil contains a host of nitrogen-fixing bacteria, which were thought to be the final link in the chain of metabolic processes in the soil. According to the general formula

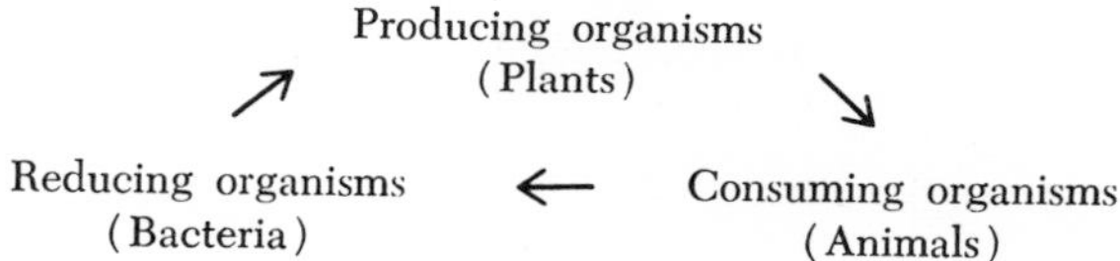

plants produce organic substances, animals eat and digest them, and bacteria reduce animal excrement and remains to the basic materials needed by plants.

There remained, however, one unsolved link in this chain—the origins and nature of humus.

Nature and Origins of Humus

The preceding "equation" is incomplete, for it fails to account for one unknown that can assume, moreover, considerable proportions. We need only think of the floor of a pine forest, where needles and dead wood often form a layer several inches deep. This layer is called "raw humus." If we dig through it, we come upon humus proper, a crumbly black mass. Seen under the microscope the individual "crumbs" have a highly symmetrical form—they are generally cylindrical, oval, or barrel-shaped. They vary greatly in size, as well. Microscopic sections of the soil show further that humus particles consist chiefly of plant material in various states of composition and often mixed with mineral matter.

Today we know that these crumbs are the excrement, or casts, of small soil animals, many of which can be identified by their "droppings" (Fig. 16). Hence the breakdown of plant matter in the soil does not proceed directly from plants to reducing organisms, but is delayed in the so-called humification process (formation of

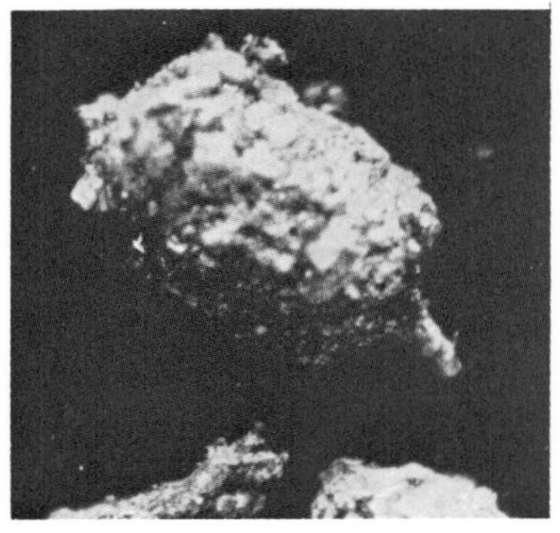

a

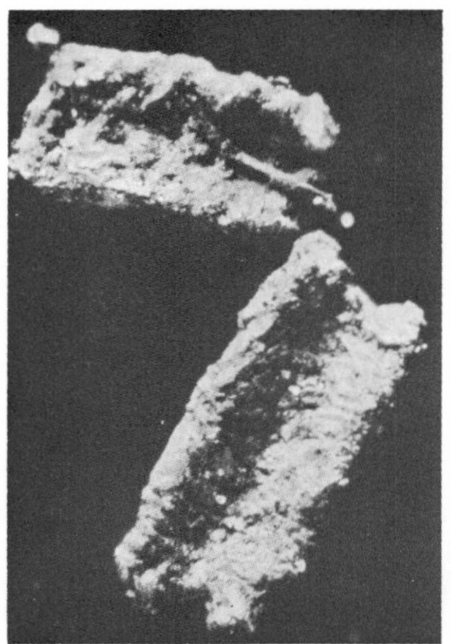

b

FIG. 16. The excrement of various soil animals. (a) Millipede (*Glomeris*). (b) Wood louse (isopod).

humus), a delay no doubt due to the excrement's being broken down in stages. It follows that humification is a partly zoogenic process, that is, it is affected by animal activity.

Chemists are still unable to tell what precisely humus consists of. All they know is that it is chemically heterogeneous, a mixture rich in lignin, which is particularly indigestible. Yet no matter whether organic debris is broken down directly by bacteria, or first eaten and "digested" by animals, the result is invariably the formation of a darkening substance which, particularly in the second instance, appears to have a crumbly structure. How important this structure is for the fertility of the soil needs no special emphasis.

Thus, bacteria and animals jointly break down organic substances—particularly plant debris—in the soil, cooperating in the following manner:

1) Most soil animals eat only those parts of plants that have already been broken down by bacteria (a process that, under sufficiently moist conditions, takes place very quickly).

2) The bacteria continue their development in the intestines of soil animals, with the result that animal excrement is often far richer in bacteria than was the original debris. According to Heymons (1923), for instance, 1 gm of dry soil from a clover field contains 11,000,000 bacteria, 1 gm of dry matter in the intestine of earthworms contains 10,000,000 bacteria, and 1 gm of dry earthworm excrement contains 52,000,000 bacteria.

3) The relatively loose excrement of the generally larger, "primary" breakdown organisms is redigested by progressively smaller soil animals, so that the humus crumbs become finer and finer.

4) Earthworms and enchytraeid worms (potworms) devour the more or less decomposed organic "soil-constituents" together with mineral matter, combining them into the argillaceous (clayey) humus we have mentioned, and which we know to be of the greatest importance to plant growth.

In the typical situation, therefore, the formation of humus is not a single, unified process, but one that takes place by stages. As a result, the composition of humus varies with climatic, phyto-sociological, and general soil condition (Figs. 17a and 17b).

Fallen leaf

Holes eaten in leaf (fenestration)
Leaf epidermis opened to minute plants (microflora) by larger springtails and bark lice

Fenestration and perforation by smaller dipterous larvae

Perforation and deskeletonization by mollusks, wood lice, millipedes, earwigs, larger dipterous larvae, larger oribatids

Maximum bacterial decomposition following increase in surface area; leaf surface eaten by enchytraeids, small springtails, and oribatids

Absorption of decomposing mass into the soil, commixture with minerals, formation of clayey humus by action of various earthworms

Repeated absorption by soil, further formation of clayey humus by various earthworms and enchytraeids

Continuous loosening of material and formation of "crumbs" by all burrowing and digging animals

Mull

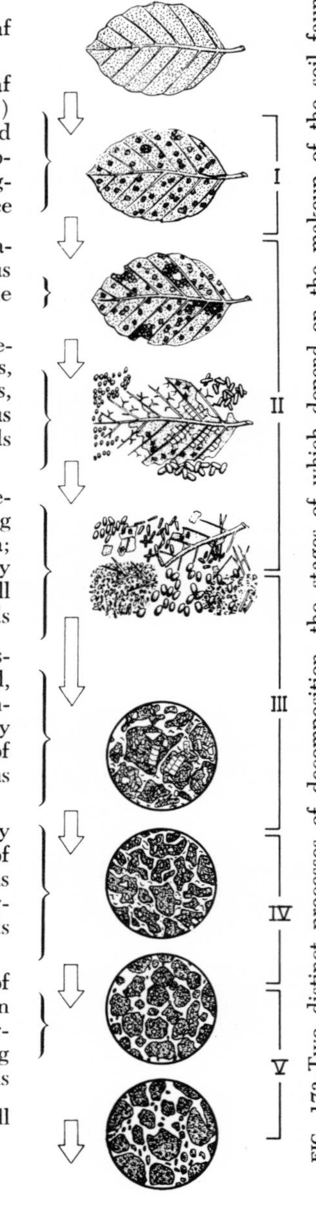

FIG. 17a. Two distinct processes of decomposition, the stages of which depend on the makeup of the soil fauna (from Zachariae, 1959). Soil animals taking part in humification can be divided into groups. Their activity on the forest floor proceeds along the following successive stages: I. Attack by animals inhabiting the litter layer. II. Comminution or pulverizing of leaf litter by arthropods and mollusks. III. Transfer of litter into humus layer by earthworms. IV. Further development of material by earthworms and enchytraeid worms. V. Formation of humus particles by the movement of soil animals. (a) Soil rich in species and individuals, where the entire leaf litter passes

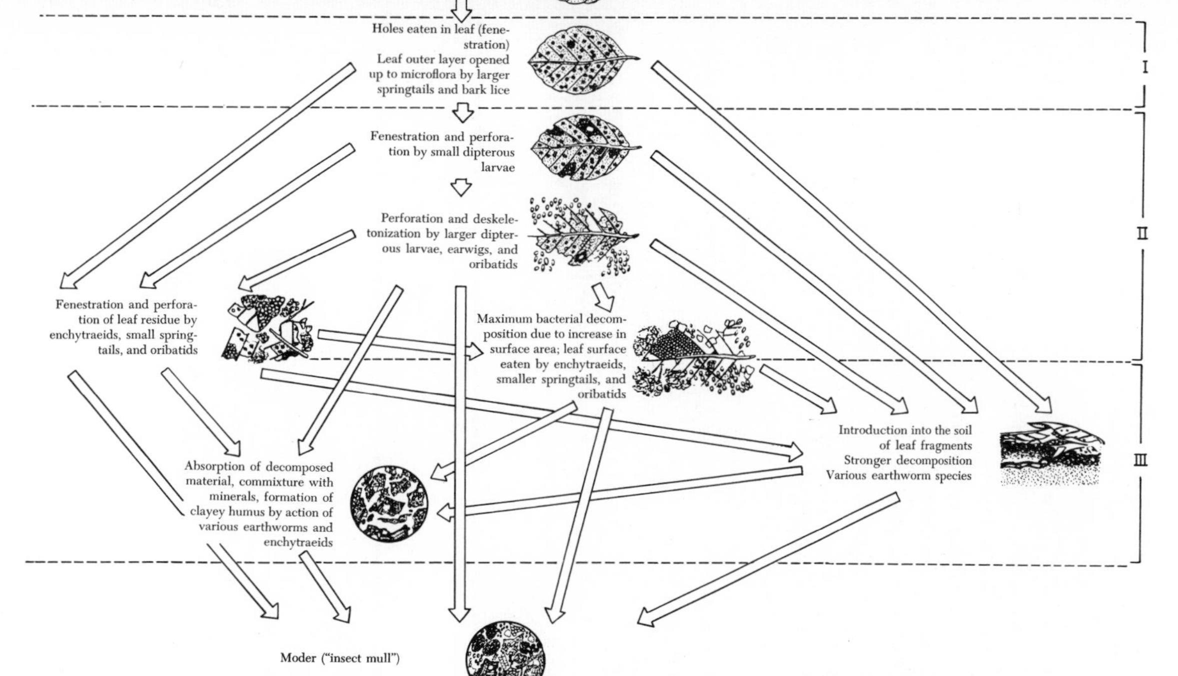

FIG. 17b. Soil with less active animals. Only part of the leaf litter passes through each of the various stages, so that the resulting humus is a mixture of organic remains in various stages of composition: moder ("insect mull") formation in beech forest.

For our purposes, however, all that matters is that soil animals form an essential link in the process of reduction or decomposition. This is most apparent in soils devoid of animals or in which the soil fauna cannot flourish, as for instance in soils that are too dry or too wet. Here, we obtain thick layers of partly undecomposed or half-decomposed litter. The resulting (raw) humus is generally sour and unusually rich in fungi. In general, there is a kind of competition between soil fungi and soil animals. Wherever the former gain the upper hand, humification is slowed down, with the result that the plant residue is not fully broken down. This also happens when the plant residue does not provide the local fauna with adequate nourishment, as in pine or oak forests. Two factors in particular are decisive in this respect: 1) the relative proportion of carbon (C) and nitrogen (N) of the food substances, and 2) their resin and tannin content. From Table 2 (Kühnelt, 1950, from Wittich, 1943) we can see that the C:N ratio of various plant fragments determines their decomposition.

TABLE 2

Tree	*Period of decomposition in years*	*C:N ratio of foliage*
Alder	1	15:1
Ash	1	21:1
Elm	1	28:1
Black elder	1½	22:1
Hornbeam	1½	23:1
Linden	2	37:1
Maple	2	52:1
Oak	2½	47:1
Birch	2½	50:1
Aspen	2½	63:1
Spruce	3	48:1
Beech	3	51:1
Red oak	3	53:1
Pine	3	66:1
Douglas fir	3	77:1
Larch	3	113:1

Moreover, many animals with a calcareous shell depend upon taking up sufficient calcium in their food. This is particularly true of diplopods, such as millipedes, which digest leaves at a rate that varies with their calcium content.

Needless to say, the process of humification of big pieces of wood—tree stumps and branches, for example —takes longer still; here the decisive factor is the relative humidity of the soil. We distinguish between *dry decomposition*, due mainly to the activity of wood-destructive insects; *white rot*, which occurs with high humidity and ends in marked fungus development; and *red rot*, which occurs in moist sites as well, but which goes hand in hand with the anaerobic (without free oxygen) development of bacteria and nematode worms. In Central Europe the humification of a tree trunk may take from 12 to 15 years (Fig. 18). The kinds of animals that participate in this process also determine the properties of the humus. Particularly in dry soils, the animals are the biosynthetic factors par excellence, since their intestines provide the bacteria with the necessary moisture. Moreover, by the mechanical comminution, or reduction, of the organic base of their excrements, they often create the conditions under which alone the bacteria can proceed to their work of decomposition.

Soil Animals and Humification

While there is little doubt about the biosynthetic importance of soil animals, it is extremely difficult to measure their contribution of chemical compounds quantitatively, for before that can be done, the following factors must all be determined:

1) The actual number of soil animals in a given type of soil.

2) Their rate of development and longevity (mass exchange).

3) Their relative size and absolute mass.

a

b

FIG. 18. Breakdown of beech stumps (unpublished drawings from original photographs; from A. Brauns, 1954). (a) 3-year-old beech stump. (b) 6-year-old beech stump. (c) 9-11-year-old beech stump. (d) 12-year-old beech stump.

4) Their metabolic activity (food requirements, time of digestion, nature and intensity of catabolic processes in the intestine, quantity of excrement).

5) The local climate and its varying influence on different metabolic types.

6) The changing relationship between herbivorous and carnivorous soil animals.

7) Possible migrations.

We are still unable to determine these and other factors with sufficient accuracy to arrive at any kind of reliable conclusion about the precise role of animals in the formation of humus. Nevertheless, since the number of soil animals in various soils (see Tables 3, 4, and 5) is very large indeed, they must play an enormous part in its development cycle.

TABLE 3. Average number of individuals found in natural forest and meadow soils (After Gsin, 1948)

	Individuals per cubic decimeter
Protozoa (Amoebae, Flagellates, Ciliates)	1,000,000,000
Rotifers and tardigrades	500
Nematodes (roundworms)	30,000
Springtails (Collembola)	1,000
Mites (Acarina)	2,000
Other arthropods (small spiders, crustaceans, millipedes, and insects)	100
Enchytraeid worms	50
Earthworms	2

However, mere census counts tell little about the metabolism of the various soil animals, the more so as they differ in size, mobility, and general activity, and also in distribution. For that reason many students have gone to the time-consuming trouble of determining the so-called *biomass* of different soils. To do this, they not only counted but also measured and weighed all the animals involved.

TABLE 4. Weight (down to 15 cm) and number of soil organisms in a Swiss meadow
(From data by Stöckli and Koffman, in Kühnelt, 1958)

	Grams per square m	*Number*
Bacteria, actinomycetes, fungi, algae	2,021.9	1.5×10^{13}
Earthworms	400.0	200 – 300
Arthropods (other than named types) and mollusks	79.9	$1 \times 10^{4} - 1.5 \times 10^{4}$
Protozoa	37.9	$1 \times 10^{10} - 1.5 \times 10^{10}$
Nematodes (roundworms)	5.0	$4.5 \times 10^{6} - 8 \times 10^{6}$
Enchytraeid worms	1.5	$7.5 \times 10^{3} - 5 \times 10^{4}$
Mites, Collembola, Protura, Campodeids	1.5	$3 \times 10^{5} - 4.5 \times 10^{5}$

TABLE 5. Population in Alpine meadows per square meter down to 10 cm
(After H. Franz, 1941)

600,000	nematodes (roundworms)
500,000	rotifers
300,000	tardigrades
10,000	enchytraeid worms
50,000	mites
50,000	springtails

These investigators discovered a correlation between distribution density and body size: the smaller a given group of soil animals is the more closely crowded together the individuals are and the greater their relative share in the total population (dominance) and their abundance. Abundance and body size are regularly in inverse proportion to each other and can be arranged in a characteristic "pyramid" (cf. Table 7 and Fig. 19).

Tables 4, 6, and 9 are examples of attempts to compute the biomass of the soil fauna. These figures, too, must be treated with circumspection, the more so as different au-

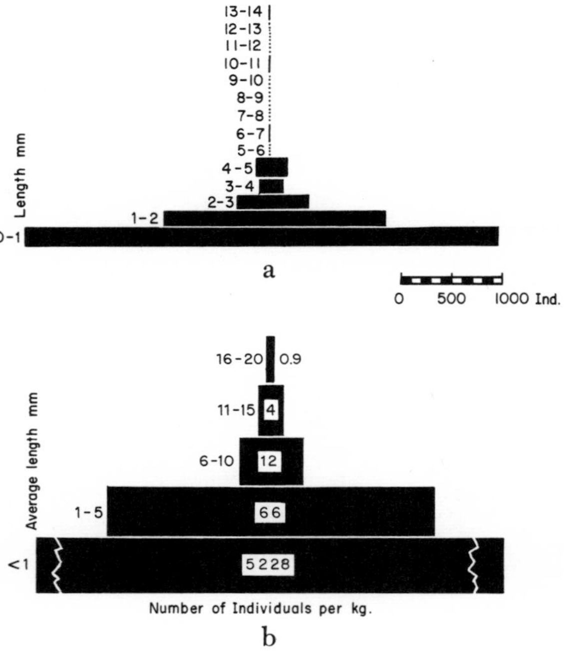

FIG. 19. Two typical "number pyramids" of soil animals: (a) An Illinois forest (from Park *et al*, 1939, in Allee *et al*, 1949); (b) tropical forest on Barro Colorado Island (from Williams, 1941, in Allee *et al*, 1949).

thors use different units of soil volumes and area, with the result that their figures are not comparable. Moreover, it is impossible to apply one practical standard to all samples, since no type of soil and no testing site resembles any other. Above all, differences in vertical distribution are essential factors that must be taken into consideration, as we shall see in the discussion of tropical soils.

The most satisfactory method is to determine the abundance of soil animals per 1000 cc of soil, if possible bearing in mind the number of species, the size of individuals (mass), the vertical distribution. After all, it does make some difference whether 50,000 individuals belong to 3 or to 30 different species, whether they are 0.5 or 3 mm in average length, and whether they congregate chiefly in the surface layer or are found throughout the entire depth of the area under investigation. In addition, the structure of the ground is of great importance, and quite particularly its so-called pore volume—the number, situation, and extension of empty spaces in the soil.

TABLE 6. Biomass of soil animals in grams per square meter: Palatinate (Bavarian) forests (From Volz in Kühnelt, 1958)

	Earth-worms	*Cock-chafer larvae*	*Round-worms*	*Small arthropods*	*Amoebas*
Beech Forest					
Total	2.015	8.605	4.052	1.817	0.873
Foliage	0.075		0.048	0.935	0.010
Raw humus (1-2 cm)	1.320		0.535	0.647	0.297
Mineral layer (to 25 cm)	0.620	8.605	3.469	0.235	0.566
Oak Forest					
Total	28.795		15.154	2.769	0.583
Foliage	0.432		0.062	0.182	0.032
Loose surface layer	1.873		0.352	1.177	0.021
Mull layer (down to 25 cm)	26.490		14.740	1.350	0.530

TABLE 7. "Number Pyramid" of soil animals in a cultivated field (Based on Stöckli, 1950, in Balogh, 1958)

Animal group and average size of individuals		*Number of individuals per cubic decimeter*
Protozoa	0.01 – 0.3 mm	1,551,000,000
Nematodes	0.3 – 1.0 mm	50,000
Small arthropods	1.0 – 5.0 mm	370
Large arthropods	1.0 – 3.0 cm	24
Earthworms	5.0 – 20.0 cm	2

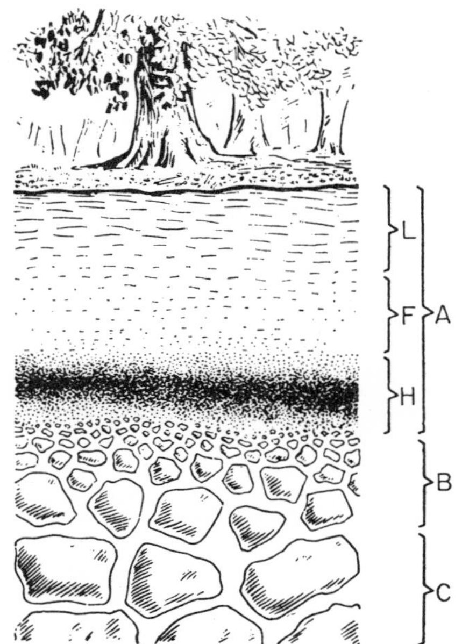

FIG. 20. Soil horizons (from Kühnelt, 1950, and Balogh, 1958): A, B, C = horizons; L = litter; F = fenestration subhorizon (raw humus); H = humic subhorizon.

Soil Horizons

A prerequisite for collecting these and similar data is a suitable method for making soil tests. Above all, a careful distinction must be made between the various soil horizons (Fig. 20). Soil biologists speak of three horizons:

A = the characteristic humus horizon, subdivided into:

A_1 = litter, with humus derived predominantly from leaves, needles, branches, moss, and grass;

A_2 = *F-layer* or fermentation layer, which consists of more strongly decomposed organic matter (may be raw humus);

A_3 = *H-layer* or humic substance layer, the zone of strongest decomposition and humification, in which only few plant residues are recognizable, while there is intimate mixing of humic substances with the mineral constituents of the soil.

B = Intermediate horizon built up by deep-reaching chemical weathering of the mineral complexes in the soil and relatively devoid of organic material.
C = Rock horizon (stones, sand, clay).

Clearly, the A-horizon is the favorite living space of most soil animals and where the most decisive biosynthetic processes take place (Fig. 21). Only the larger soil

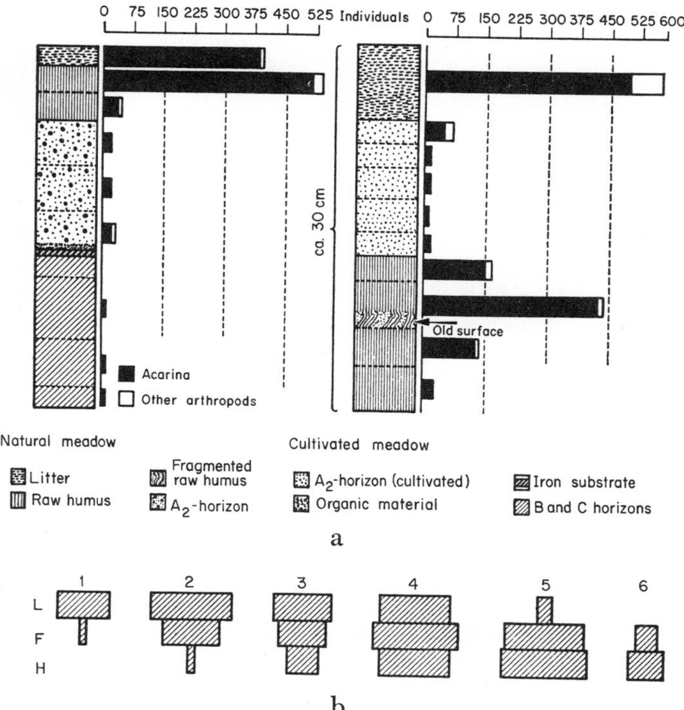

FIG. 21. (a) Vertical distribution of small fauna in a natural and cultivated meadow in England (from Murphy, 1953, in Balogh, 1958). (b) Vertical distribution of mites and springtails in the A-horizon (from Forsslund, 1943). L = litter, F = fermentation layer, H = humic layer. 1. *Phthiracarus piger*, 2. *Platynothrus peltifer*, 3. *Isotoma notabilis*, 4. *Oppia translamellata*, 5. *Ceratozetes hesselmani*, 6. *Suctobelba similis*.

inhabitants, particularly the earthworms, regularly cross the various horizons and layers, thus playing an essential part in the transport and commixing of materials. However, many smaller animals are capable of crossing from layer to layer, though often during certain seasons only.

There are, incidentally, several types of soils without a B-horizon, and in these the humus is found immediately on top of the rock. Such types include the well-known black earths (chernozems) or the so-called rendzinas, which often occur as typical "primary soils" on a firm substrate. Under "normal" climatic conditions, the soil will "grow" from the bottom upward and from the top downward—the weathering of the C-horizon and the comminution and decomposition of the organic surface layer combine their actions.

A special problem is the reaction of soil animals to agricultural intervention, by which the natural layers of the soil are clearly disturbed, and by which the smaller soil inhabitants are greatly affected. Thus, the number of individuals found in tilled fields is often no more than one-tenth to one-fifth of the number present in uncultivated meadows (cf. Tables 8 and 9).

On the other hand, the rapid rise in population in a field that had been treated with CS_2 (carbon disulphide) shows just how quickly some sections of the soil fauna can recover from catastrophes (cf. Table 10).

The natural chain of events which in cultivated land

TABLE 8. Reduction of mesofauna* in cultivated land (From Tischler, 1955, in Balogh, 1958)

	Individuals per square meter		*Ratio*
	1) *Fields*	2) *Meadows*	1):2)
Nematodes	2,000,000	10,000,000	1:5
Mites	30,000	180,000	1:6
Springtails	15,000	90,000	1:6
Enchytraeids	4,000	40,000	1:10

*Greek *mesos* = medium, i.e., medium sized.

TABLE 9. Reduction in the number and weight of individual earthworms in cultivated land (cf. Fig. 22) (From Zicsi, 1957, in Balogh, 1958)

	Earthworms per square meter					
	Adults		*Juveniles*		*Total*	
Vegetation	*No. of individuals*	*Weight grams*	*No. of individuals*	*Weight grams*	*No. of individuals*	*Weight grams*
Meadow (permanent)	89.70	96.68	82.10	38.69	171.80	135.37
Lucerne	19.25	13.22	37.95	7.96	57.20	21.18
Sugar beet	16.15	11.64	30.80	6.84	46.95	18.48

is broken by land clearance, single-crop cultivation, and regular harvesting can be closed again with fertilizers; moreover, man can substitute the plow for the earth-loosening activity of soil animals.

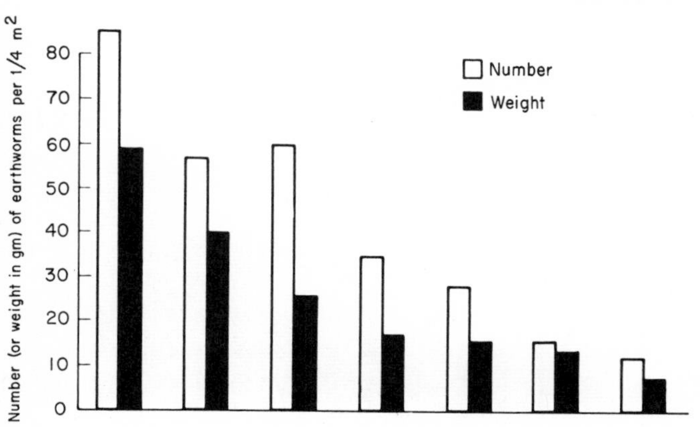

FIG. 22. Reduction of earthworm population in cultivated fields (from Tischler, 1955).

TABLE 10. Effect of CS_2 (400 gm per square meter applied on May 12, 1956) on the distribution of soil arthropods in a Palatinate (Bavarian) vineyard (After W. Huther, 1961)

Year	*1957*											*1958*		*Total*
Month	*Feb*	*Mar.*	*Apr.*	*May*	*June*	*July*	*Aug.*	*Sept.*	*Oct.*	*Nov.*	*Dec.*	*Jan.*	*Feb.*	
Control field, untreated														
Springtails	61	72	225	55	76	133	189	164	219	175	118	149	84	1,720
Mites	103	113	118	63	63	139	191	133	186	187	123	139	123	1,810
Symphyla	4	3	1	1	1	3	6	8	6	1	2	3	4	43
Dipterous larvae	2	1	...	2	5	...	...	...	2	1	...	...	3	16
Millipedes	...	...	...	...	...	...	1	2	...	...	1	...	2	6
Beetles (+ larvae)	...	2	...	...	...	...	6	...	1	...	1	...	...	10
Poisoned field														
Springtails	3	15	78	513	75	47	161	103	282	508	207	147	448	2,499
Mites	6	1	12	38	167	107	112	52	86	152	45	35	85	898
Symphyla	...	...	...	...	...	...	...	...	...	...	...	...	...	...
Dipterous larvae	...	...	...	3	1	...	...	1	...	15	1	2	...	23
Millipedes	...	...	...	...	...	...	...	...	...	...	...	...	...	...
Beetles (+ larvae)	2	...	...	...	4	...	...	...	...	2	...	...	2	10

But whenever soil animals are missing or much fewer for reasons other than human intervention, their biosynthetic importance becomes strikingly obvious. A good example of this is provided by conditions in the South-American tropics.

Tropical Soils

The Amazon basin is covered with lush, dense rain forests and is apparently a land of plenty. Yet jungle clearance and plantation work produce extremely poor results: after only 3 or 4 years the plantation becomes exhausted and has to be replaced. Hence agriculture as we know it does not exist. The contrast between the lush primitive jungle and the poverty of cultivated land in it is due to soil-biological conditions. Even the densest jungle is practically devoid of humus—there is an H-layer of 2-3 cm maximum depth, covered by the thinnest of litter layers. The A-horizon (maximum thickness 5 cm) gives way, almost without transition, to a dense layer of very heavy red clay (laterite). As a result, animal life is compressed into the narrow surface region. The B-horizon is almost completely devoid of animals, and there are no earthworms. The absence of raw humus, despite the mass of falling vegetable matter, is a direct consequence of the climate, and high temperatures and high humidity ensure that bacteria and soil fungi flourish at the expense of the soil fauna. Since microorganisms do not turn vegetable litter into larger particles but reduce it directly to more soluble forms, no humus is created or deposited. The daily downpours then wash the soluble and easily suspended materials into the ground, where they are quickly absorbed by the shallow roots. The lushness of the tropical forest is due to the accelerated metabolism. The soil has no reserves. This becomes obvious the moment the cycle is interrupted at any point,

FIG. 23. Typical clearance in the upper Amazon rain forest.

as for instance by clearance, which destroys the entire plant cover, and with it most of the biomass. This is particularly true of land clearance by burning, which also leads to the destruction of the surface fauna (cf. Fig. 23).

Two observations corroborate the above assertions: 1) Microscopic sections show that tropical soils are exceptionally rich in fungi but relatively devoid of excrement from small animals; 2) on the western edge of the Amazon basin, where the rain forests give way to mist forests, the soil is much richer in humus deposits (Fig. 24). Here the average temperatures are below 20° C., and the action of bacteria and fungi on the litter is slowed down, with the result that soil animals can create a humus reserve. And, in fact, a highly developed and varied fauna inhabits these regions. Admittedly, here, too, humidity can be so high that neither microorganisms nor animals can break the litter down; in that event a spongy mass of peat is formed.

There is a clear quantitative relationship between the composition and density of the soil fauna and the formation of humus—this applies to temperate zones no less than to the tropics, though we cannot as yet make precise quantitative comparisons. We shall describe some attempts to arrive at accurate figures, but it must be stressed that "equations" with so many unknowns are not really satisfactory.

FIG. 24. Mist forest in the eastern Andes (Peru) with characteristic tree ferns.

Computation Methods

Stöckli (1950) has estimated that some 48,000 kilograms of animal excrement are produced annually per acre of cultivated land, earthworms accounting for 75 percent of this figure. Since in temperate regions this excrement neither decomposes nor becomes reduced at once, but is largely reconsumed, significant quantities of energy are stored up in the soil. He showed further that the live weight of the microflora and total fauna in the upper 15 cm of a normal agricultural field amounted to approximately 10,000 tons per acre.

Since Stöckli underestimated the density of several groups of animals, his figures are on the low side. However, he assumed that soil organisms were evenly distributed, and this has not yet been proved.

An assessment of the importance of soil animals must not, of course, be based exclusively on their number, size, and weight (biomass), but must also take into account the different metabolisms of different species. In many soil animals the relationship between body size and food consumption is related by the formula

$$\frac{c}{\sqrt[3]{g^2}} = \text{constant},$$

where c is the food consumption and g the "active" surface. In other words, the smaller an animal the more food (per unit weight and time) it needs. The criterion is not body weight, but (inner plus outer) body surface, which is relatively greater in small than in large animals (see Table 11).

Dunger (1958) has, however, emphasized that the formula differs from species to species. That the food consumed by litter-eaters depends on their specific "taste" has been mentioned. Dunger found that millipedes *(Iulus scandinavius)* and wood lice *(Ligidium hypnorum)* do not like dried up alder and maple leaves. In other words, there are clear differences in eating habits, quite independent of the consumer's size. He

TABLE 11. Relationship between body size and leaf consumption and Drift ratio in the millipede *Glomeris marginata* (from van der Drift, 1950, in Balogh, 1958)

Average body weight of individuals in milligrams $= g$	*Daily food consumption as a percentage of body weight*	*Food consumed by an individual in 5 days (in milligrams of air-dried matter)* $= c$	$\sqrt[3]{g^2}$	$\frac{c}{\sqrt[3]{g^2}}$
51.2	70	60.0	13.8	4.3
52.9	62	54.4	14.1	3.9
52.2	66	58.8	14.2	4.1
114.7	49	94.2	23.6	4.0
116.4	46	88.5	23.8	3.7
122.1	52	104.8	24.6	4.3
185.5	33	100.8	32.5	3.1
186.5	43	133.2	32.6	4.1
199.8	32	107.4	34.2	3.1

was able to show that julids eat four to eight times as many of last year's leaves as they consume fresh ones, while wood lice consume less but identical food throughout the year.

Dunger also compared the C:N series reported in Table 2 with the results of his own "preference" studies and discovered a clear relationship between rate of decomposition and food preference.

It follows that millipedes and wood lice prefer the leaves that decompose most quickly. The same is true not only of these arthropods but also of many other soil animals, including earthworms, enchytraeid worms, snails, oribatids, and springtails.

In one site examined (a mixed ash-oak forest) millipedes and wood lice consumed approximately one-third of the annual leaf fall (150 grams per square meter), the millipedes accounting for about twice as much as the wood lice. In other sites, with a considerably greater leaf fall (beech forest = 500 g per m^2, Volz, 1954), the share of these two groups is correspondingly smaller. It has been estimated that all soil animals together consume no more than 40 percent of the annual litter.

It must be reemphasized that these figures are no more than crude estimates; at best they apply to a given site and to individual species of animals. The chemical properties of the food (for instance, its C:N ratio or its cellulose, lignin, tannin, and resin content), its state of "fermentation" affected by bacterial action and humidity, the seasonal fluctuations in animal food requirements, uneven growth rates, climatic influences, and many other factors must all be taken into account.

Nevertheless, the collection and comparison of these quantitative data represent a great advance on the time when earthworms and bacteria alone were treated as essential links in the soil-biological chain. Humus research has become a biological discipline in its own right—one in which zoologists, botanists, microbiologists, chemists, and mineralogists increasingly pool their resources.

III

THE HABITS, SENSE ORGANS, AND BEHAVIOR OF SOIL ANIMALS

To the biologist, soil animals are of special interest not only for being an important link in the metabolism of the soil, but also for their own sakes. Biologists do not make the "usefulness" or "potential harmfulness" of an organism a criterion of its "importance." Zoologists are used to the jibe that they worry about the soul of the cockchafer. In what follows we shall try to show that this concern is extremely rewarding—both in theory and also in practice.

Tactile and Olfactory Organs

Life in the earth is a kind of troglodyte existence which is reflected in the sense organs and reactions of soil animals. Touch and smell are their most important means of orientation. This becomes clear from an examination of the corresponding organs, which are often inordinately large and highly differentiated. As an example we need only think of the pseudostigmatic organs of beetle mites (oribatids); these tactile hairs of various shapes and sizes are embedded in special pits in the integument and supplied with special nerves. They may be straight or arched, smooth, feathery, or club-shaped, but whatever their shape they can detect the slightest vibration in the soil or the air. In beetle mites, stimulation of these tactile hairs causes the animals to feign

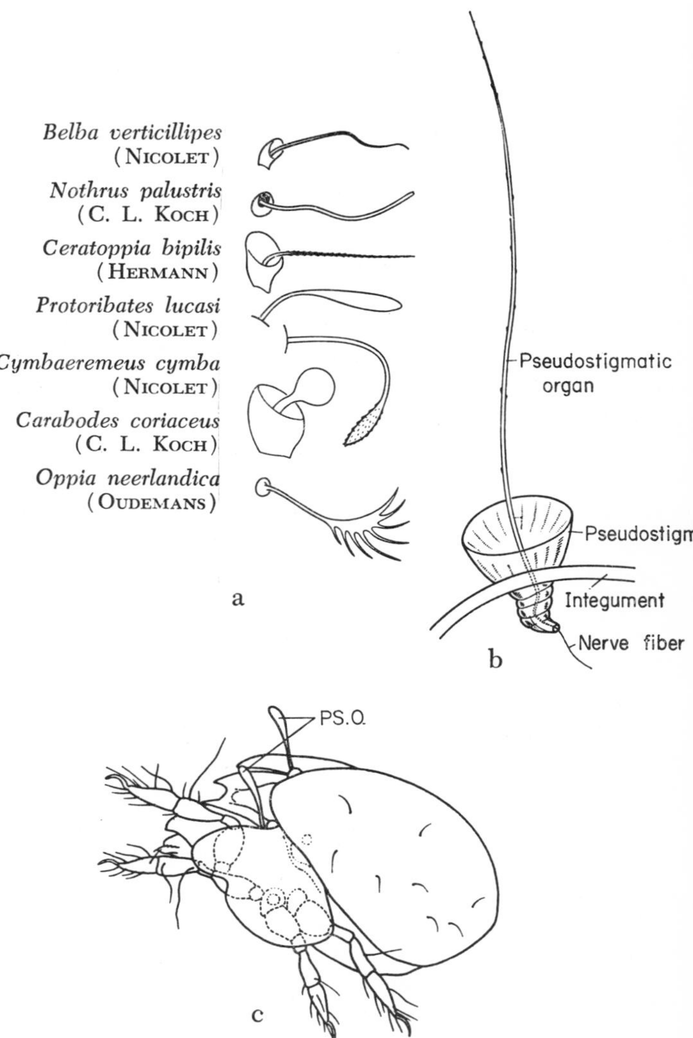

FIG. 25. (a) Tactile hairs (pseudostigmatic organs) of various beetle mites (oribatids); (b) tactile hairs of beetle mite (*Belba geniculosa*); (c) position of tactile hairs in the *Epactozetes* mite (from L. Beck).

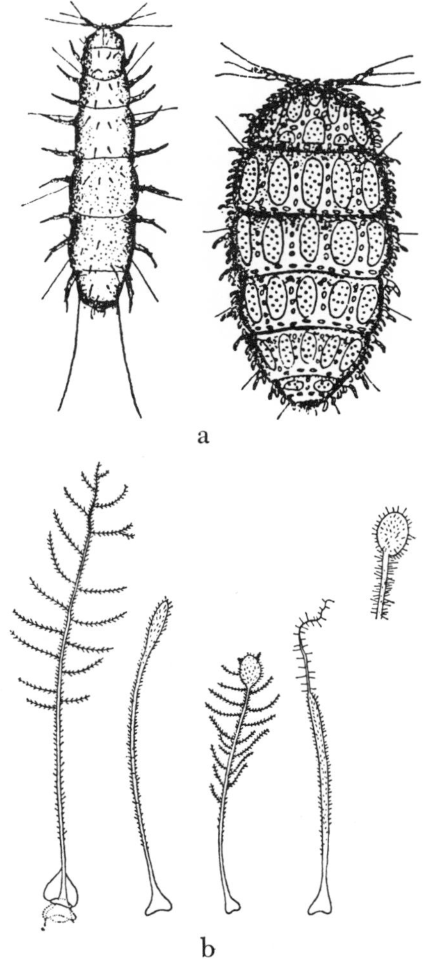

FIG. 26. (a) Pauropods —myriopods measuring several mm (from Kühnelt, 1950). (b) Various types of sensory hairs in pauropods (from Delamare-Deboutteville, 1951).

dead at once. However, it took zoologists a long time to discover the precise function of these hairs, so much so that five distinct hypotheses about them were put forward. They were variously described as breathing organs, olfactory organs, ears, moisture receptors, and temperature sense organs.

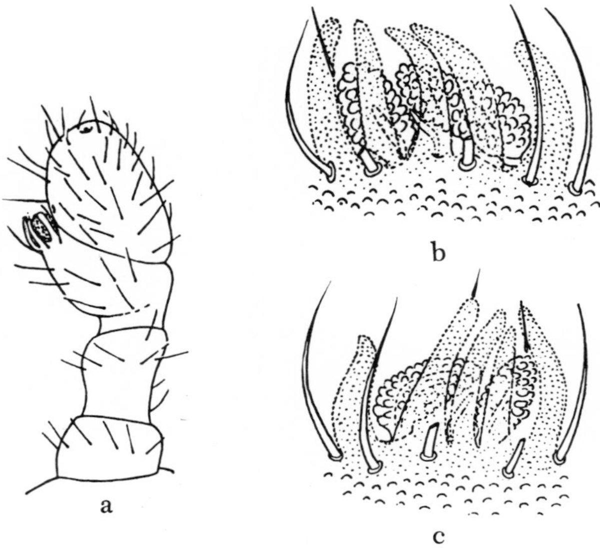

FIG. 27. Antennal organs of soil-dwelling springtails. (a) The reduced antennae of a springtail (*Onychiurus*) bearing prominent bristles, cones, and grapelike appendages on its third segment. (b) and (c) The same organs greatly magnified.

Still, mere hypotheses were not enough; what scientists do when in doubt is to devise suitable control experiments. In a long series of investigations, Pauly (1956) accordingly compared the reactions of intact beetle mites (of the genus *Belba*) with those of beetle mites the tactile hairs of which had either been removed or sealed up, and he thus discovered the true function of these organs. Since no winds disturb them in their natural habitat, the wind-shy (anemophobic) behavior of beetle mites makes good sense: it prevents them from straying. This probably explains why in aquatic oribatids these hairs are particularly short and stub-like (Fig. 25).

Such tactile body hairs (or trichobothria as biologists also call them) are widespread among other soil-dwelling arthropods as well—they are present in a host of millipedes, primitive insects, and spiders, though they are rarely as prominent as in beetle mites (cf. Fig. 26).

In addition to these tactile organs, many soil animals also have highly developed organs of smell (olfactory organs or chemo-receptors). A good example is provided by the springtails, of which we have already met a number of soil-dwelling forms. While surface inhabitants among them have elongated but "normally" constructed antennae, depth dwellers have shorter antennae bearing prominent organs, probably associated with the sense of smell (cf. Fig. 27). Biologists are still trying to discover their precise function. In many soil animals, moreover, various mouth parts have been developed as specific tactile and olfactory receptors. This is particularly true of spiders, all of which lack antennae.

Humidity, Temperature, and Acids

Since the relative humidity of the air plays so important a role in the life of soil animals, it is only to be expected that they should be particularly sensitive to changes in it. It can readily be shown that the small forms need a great deal of moisture in the air—at normal room temperature and humidity they wither within a matter of minutes. But even the larger and more robust species depend upon humidity; thus, if wood lice are placed in a receptacle with a moisture gradient, they will quickly make for the moister side. As long as the air in their environment is too dry, they remain restless. This stimulating effect of humidity, known as hygrokinesis, continues until the animal discovers optimum moisture conditions. For soil animals in general, the optimum is above 90 percent.

Most small soil dwellers have a water-permeable skin, and as a result they lose a great deal of body fluid by evaporation in dry air. When free liquid is available they can make up the loss by drinking, but drinking water is not generally available in the soil, except after rains; at other times the moisture is bound to the soil particles themselves or by capillary action between them. However, as water in the interstices of the soil keeps evaporating, the air in it is generally saturated. Consequently, true soil dwellers, which as we saw lack effective anti-evaporation mechanisms, are physiologically restricted to their "cave" atmosphere. The moisture sense of insects, incidentally, is localized in their antennae.

What we have said about the moisture sense applies equally to the temperature sense. In a container in which the temperature is graduated, soil animals invariably make for regions of lower temperature and try to avoid all strong fluctuations. Their temperature optimum (in moderate climates) is below 20°C, and generally around 15°C.

The atmosphere of the soil has other special characteristics to all of which soil animals have become adapted. To begin with, the proportion of carbonic acid in the soil may be considerably higher than it is in the atmosphere. While all higher animals are extremely sensitive to an excess of carbonic acid, small soil dwellers are far less so. Springtails can be kept for hours in an atmosphere containing 20 percent of CO_2, to awaken without damage after a short stupor. Earthworms have glandular swellings in their narrow esophagi which secrete calcium and also store it for some time. These swellings are known as the calciferous glands, and many authorities associate them with CO_2 resistance, since calcium readily combines with carbon dioxide. It seems likely that excess carbonic acid in the blood of earthworms is combined with the calcium in these glands and is excreted as chalk by the intestine.

Other authorities attribute another, though similar, function to the calciferous glands; according to them these glands supply the calcium for the neutralization of the various humic acids which earthworms ingest with the rest of their food.

The acid or base content of the soil does not seem to influence the soil fauna overmuch, and within certain limits the distribution of species is hardly affected by it. This is particularly true of the small arthropods. Only those having calcium shells—such as snails, wood lice, and millipedes—are more obviously restricted to calcareous soils.

Earthworms Between Scylla and Charybdis

That earthworms so often come up from the ground after a heavy rainfall, only to perish on paths and in puddles, is largely connected with a respiratory problem—the water threatens to asphyxiate them in their flooded burrows. Once they emerge from the ground, however, they quickly die of exposure to light, for they are extremely sensitive to ultraviolet rays, which are present even in diffuse daylight. Their situation is comparable to that of the Greek sailor caught between Scylla and Charybdis—by emerging from the flooded burrows into daylight they are leaving the frying pan for the fire.

Even the giant Australian earthworms *(Megascolides australis)*, which can attain lengths of more than 8 feet, are completely helpless outside their burrows. They are extremely difficult to catch, and it takes a strong man to pull one out of its lair (Fig. 28). The aborigines consider them a great delicacy.

Sensitivity to Light

Though the eyes and light receptors of most soil animals are poorly developed, even blind forms have light-sensitive skins. After only the briefest stay in the light, they become restless and begin to move about until they find a dark spot. Earthworms are especially light-sensitive in their extremities, particularly in the front end, and if light is thrown upon it they invariably

FIG. 28. Australian aborigenes hunting giant earthworms. These animals throw up craters at the top of their burrows.

beat a hasty retreat. The rest of their skin, too, is studded with light receptors. Under normal conditions their photophobic behavior goes hand in hand with thigmotactic reactions—the desire to be touched on all sides—which, with them, manifests itself in a tendency to crawl into narrow cracks and gaps in the earth. This makes it exceptionally difficult to study them. It is possible, however, to neutralize both their light-shunning behavior and their love of contact with their surroundings by placing them under panes of glass and observing them in moderate light.

Observing Habits

This brings us to the problem of the best ways of studying the habits and reactions of soil animals in general. All such studies presuppose that we can keep the experimental animals alive and, if possible, breed them as well. It is only to be expected that they should be extremely sensitive and demanding experimental material, coming as they do from a habitat with an exceptionally even climate. Throughout the whole year, conditions of life, such as moisture and temperature, fluctuate much less beneath the vegetative cover than they do at the surface. Even such factors as the food supply are remarkably constant. Hence if we wish to keep and breed soil animals under laboratory conditions, we must first of all offer them an even environment. This is now done by means of climatic chambers. But even without these, soil animals can be bred on a small scale, provided only that we avoid marked temperature and moisture fluctuations in the containers and suppress fungal growth. This applies particularly to the smaller species, which suffer most from water condensation and mold formation. A much more difficult task is coping with their light-shunning behavior and the associated tendency to crawl away from the observer. They can be confined in special dishes with flat glass bottoms so that they are unable to hide, but in that situation they no longer display normal behavior, not even after long periods of "training."

For some years biologists have been using the following techniques:

1) Small and medium-sized animals are placed in small glass cylinders with smooth plaster bottoms. These can be kept evenly moist, are easily cleaned, and help to keep the animals in view.

2) Newly introduced animals are continuously exposed to weak illumination, which does not really harm them, even though they normally avoid it. After some

FIG. 29. Containers for rearing small soil animals. (a) Glass cylinder. (b) Plaster plate with special hollows covered with glass. The central grooves are filled with water to ensure even moisture conditions.

a

b

time they become used to it and behave in weak light as they would normally behave in the dark.

3) Species showing their desire to be in tactile contact with their surroundings can be placed between loose glass platelets, or better still in shallow plaster dishes covered with glass (Fig. 29).

Continuous light is needed because very small animals can only be studied under a lens. If the animals are left to the normal alternation between day and night, they will become nocturnal in habit and escape from microscopic observations during the day or in artificial light.

Because the animals are so highly sensitive to air and ground vibrations, the vessels must remain covered with glass even during observation, and the microscope must be kept out of direct contact with the glass or, better still, in free suspension.

Feeding Habits of Vegetarian Soil Animals

Having made all these preparations, we can observe at leisure the comings and goings of these small creatures. We may catch a glimpse of one of them eating its food; more likely than not it will be a vegetarian, for the food of herbivores does not run away and can be consumed slowly. To provide the right kind of diet is not difficult, since soil animals flourish on what the ground has to offer: leaves, bits of wood, fungi, roots, and moss. The favorite food is also the most common: leaves that have been "softened" by water and bacteria (cf. Fig. 30). Vegetarian millipedes, mites, springtails, bristletails, and others first eat holes into the leaves and then skeletonize them down to the ribs, or else remove only the fleshy parts inside the leaf, to produce the kind of patterns shown on Fig. 31. Others, including oribatids, prefer to eat bits of decaying wood; still others—for instance, springtails, some of the Protura (primitive insects whose first pair of legs serves as antennae), or a number of Pauropoda (some of the smallest millipedes)—feed on fungi or suck up nourishment from the threads (hyphae) of a mycorhiza (an interwoven mass of these threads and root tips forming a symbiotic association between a fungus and a higher plant). Only a very few eat moss; many feed on excrement and help to break it down further, among these are the very smallest—springtails, mites, symphyla arthropods—which can therefore be considered as secondary decomposers.

FIG. 30. Earthworm pulling pine needles into the soil. The animal is burrowing between two plates of glass. The needles become softened in the soil; on the soil surface they are too hard for the worm to bite through.

Predators and Their Methods of Attack and Capture

Much more exciting are the feeding habits of predators, which involve searching, lying in wait, pursuing, attacking, seizing, killing, dismembering, and finally eating.

Those predators inhabiting the upper layers of the soil—the leaf litter—include some that detect their prey

FIG. 31. Typical marks left by (a) springtails on hornbeam leaf. (b) Larvae of *Penthetria holosericea* on black poplar leaf (from A. Brauns, 1954).

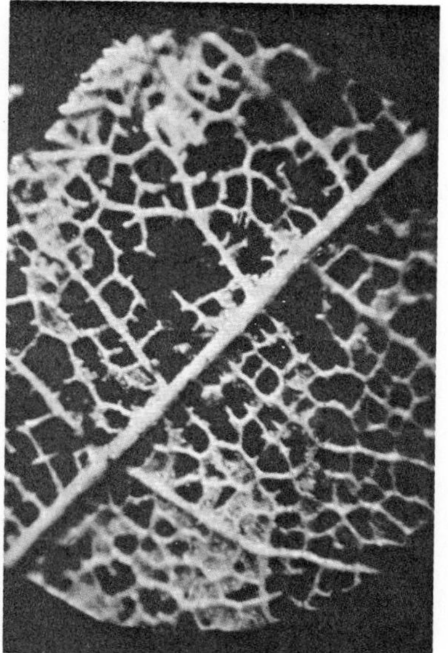

a

b

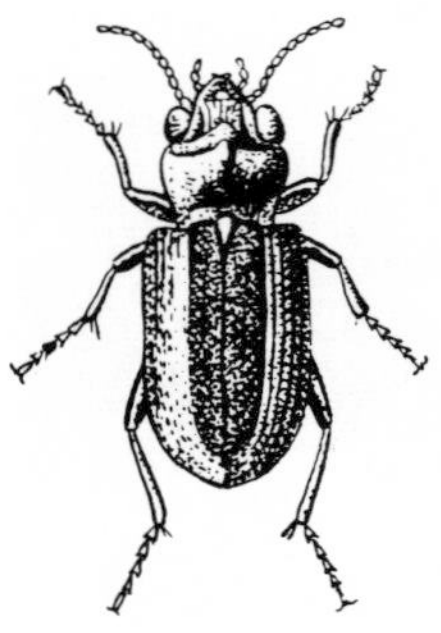

FIG. 32. A small carabid beetle (*Notiophilus biguttatus*).

by optical means. An example is the common carabid beetle *(Notiophilus biguttatus)*, which has large and protruding eyes (Fig. 32). Most of the time, this beetle sits perfectly still, but the moment a small animal starts to move in its vicinity, the beetle pounces on the prey with great agility and seizes it in its "fangs." Springtails are its favorite food, probably because they are soft. Hard beetle mites are generally spurned.

Most other predators have poor eyes or are totally blind, so that they must smell or feel out their prey. These animals are constantly on the move, touching everything in their path with legs, antennae, and mouthparts, and putting their "noses" into every crack. This behavior is seen best in the actions of *Japyx*, a primitive insect armed with sharp anal forceps. This insect is a true soil dweller, and it, too, prefers springtails to all other prey. It resembles the earwig in shape and can crawl into the smallest crack. Once it has spied out a springtail—of the white, blind sort which cannot jump—its movements become incredibly quick. Generally, it will seize its victim at the first attempt. If this proves impossible—if the crack is too narrow—*Japyx* will swiftly turn round, push its forceps into the crack, and seize the prey in that way. The victim is then twisted out of the crack and eaten "off the fork" (Fig. 33).

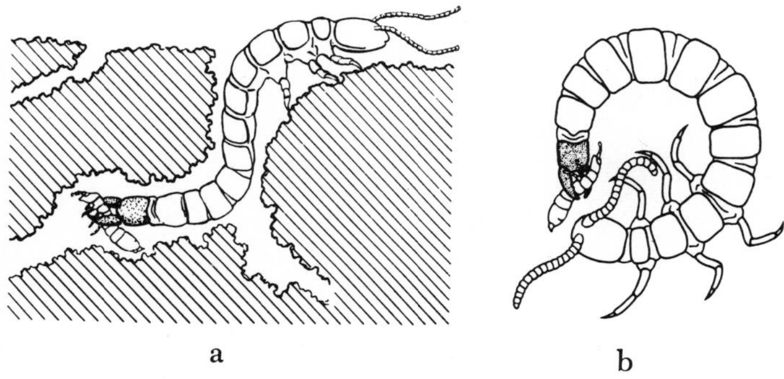

FIG. 33. A primitive insect (*Japyx*). (a) Seizing a springtail (*Onychiurus*) in its double forceps. (b) About to eat its prey.

Many predatory millipedes, such as scolopenders, detect and pursue their prey in similar ways. They, too, thrash about with their antennae and, the moment they touch an edible animal, swiftly seize it in their large and pointed jaws and kill it by the injection of a poison (Fig. 34). Millipedes seem to have an innate capacity for detecting suitable prey, and many of them specialize on certain victims only. This is particularly true of the elongated geophilid centipedes, all of which have a clear preference for worms (see Fig. 12a). The larvae of tiger beetles are among the most ingenious trappers and predators (Fig. 35).

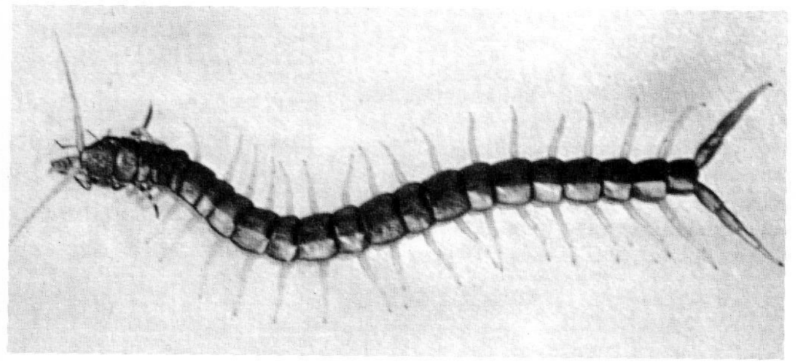

FIG. 34. Scolopender with prey (meal beetle).

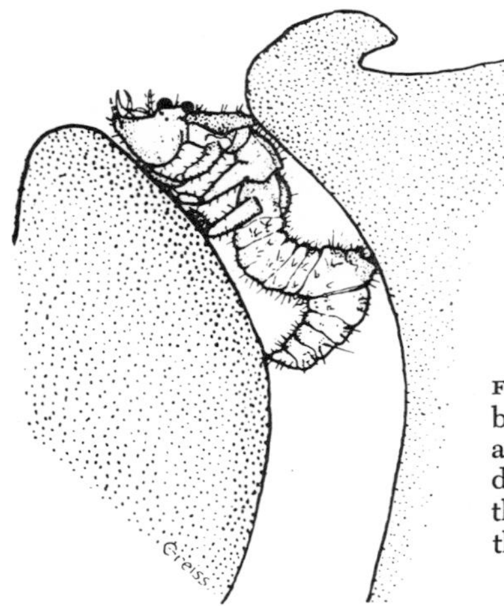

FIG. 35. The larvae of tiger beetles (*Cicindela*) lie in wait at the top of vertical cracks and drop down with their victims the moment one of them crosses the crack.

We have already mentioned that spiders and their relatives constitute the majority of predatory soil animals. Among them the pseudoscorpions are the commonest in temperate regions. They lie in hiding for their victims or walk about slowly to seize passing animals in their powerful claws. They and their southern relatives, the true scorpions, will be referred to later.

Daddy longlegs, or harvestmen, include a soil-dwelling group that lives exclusively on snails. They have an exceptionally flat and hard armor and inordinately long legs, whence their name. Once they have caught a snail in their scissor-sharp jaws and consumed it, they, or rather the females, deposit their eggs in the empty shells (Fig. 36).

Nor are daddy longlegs the only snail-lovers among soil animals. Large beetles of the genus *Carabus* are also very fond of snails. One of their near relatives, *Cychrus*, has a very pointed mouth with which it can reach into the shell of snails without first having to break them open. The larvae of the glowworm are also particularly keen snail hunters (Fig. 37).

a

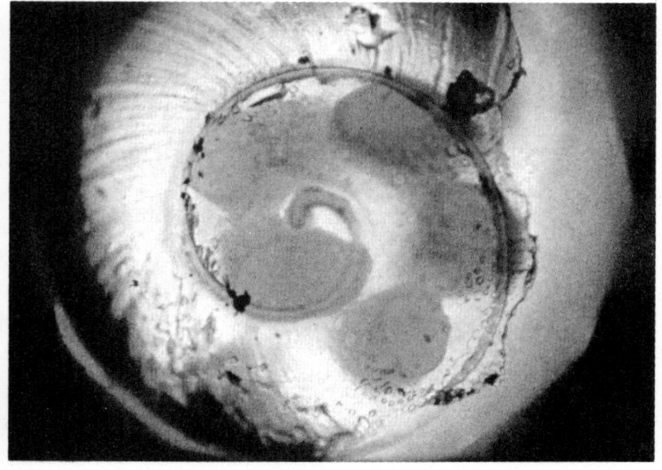

b

FIG. 36. (a) *Trogulus* individual attacking a small snail; actual size 0.8 cm. (b) *Trogulus* eggs in an empty snail shell (from W. Pabst, 1953).

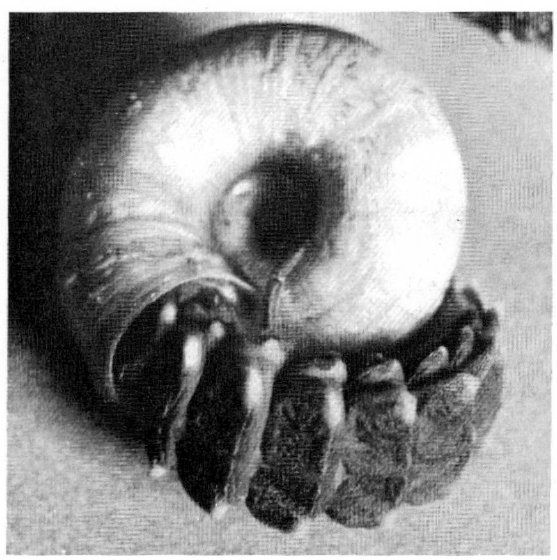

FIG. 37. Larva of glowworm (*Lampyris*) preying on a snail (from H. Schwalb, 1961).

We must mention one more predatory spider—*Scytodes thoracica*—because of the unusual method it uses to catch its prey (Fig. 38). In temperate regions it is often found in houses, but in the south it generally lives under stones or in the soil. This spider covers its victims with threads of sticky "spittle," thus pinioning them to the ground before killing and devouring them. The threads are produced from preoral appendages known as chelicerae, and Dabelow (1958) has been able to show on film that the "spittle" is directed so accurately that the prey is covered with a tight zigzag band, from which not even the agile silverfish can escape.

"Spitting" also occurs in another group of soil animals, quite unrelated to the *Scytodes* spider. This group includes *Peripatus* and its relatives (the Onychophora), all of which live in the leaf litter of moist tropical forests

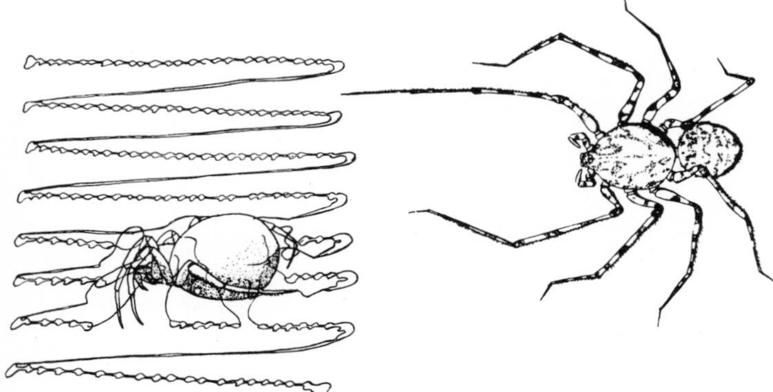

FIG. 38. A predatory spider (*Scytodes thoracica*) pinioning its prey, a small spider, to the ground with threads of sticky "spittle." The zigzag shape of the "net" is produced by rhythmic alterations in the pressure of the jaws.

(Fig. 39). These stub-legged animals are several centimeters long and look rather like a cross between a worm and a millipede. They produce a gluey spittle, somewhat less symmetrically but no less effectively than the

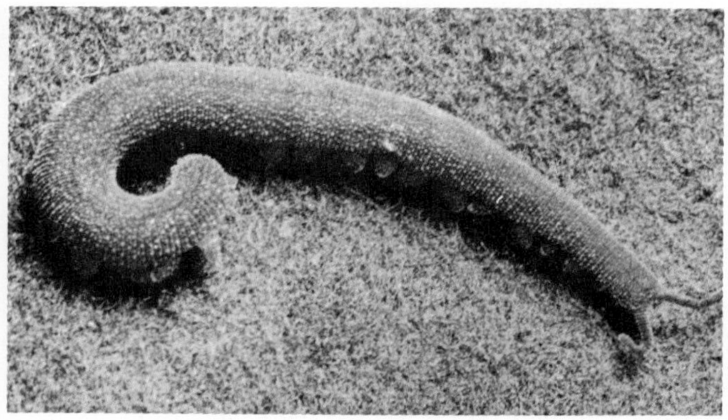

FIG. 39. A member of a strange group (*Peripatus*) that apparently occupies an intermediate position between worms and myriopods. Note the characteristic foot stumps. Members of this genus prey on worms and snails. Actual size 3-4 cm (photograph: H. Sturm).

Scytodes spider, from two special papillae on either side of the mouth. Observations suggest that with them, however, the "spittle" is not so much used for catching prey as it is for defense purposes.

Defense Mechanisms of the Prey

The victims try to elude their pursuers by a host of different ruses. Most of them simply play dead the moment they are subjected to strong stimuli. Vegetarian millipedes and springtails roll up and remain motionless (Fig. 40). Various mites and wood lice bend over forward, thus protecting their more fragile appendages and body surfaces (Fig. 41). The peak of specialization, however, is probably found in the pill bug *(Armadillidium)* and in the pill millipede *(Glomeris)*. The pill bug has perfected the folding-up technique of "normal" wood lice to such an extent that it can roll up into a ball so that its front end can barely be distinguished from its back (Fig. 42). No less perfect spheres are produced by the pill millipede which, in the expanded state, looks like an elongated wood louse, so much so that it is difficult

FIG. 40. A julid millipede playing dead.

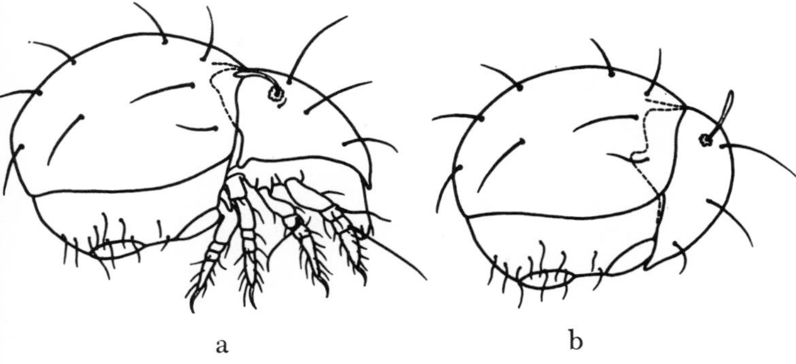

FIG. 41. A mite (*Mesoplophora*). (a) In its normal position. (b) Rolled up.

to convince the layman that it is a millipede. Once it is rolled up, even the expert is easily confused.

These two forms bear a strong resemblance to the armadillo which, as everyone knows, is a soil animal as well, but one that is completely unrelated to them (Fig. 43). Here we have a striking example of "convergence," a term zoologists use to refer to the tendency of animals living in a uniform environment to develop similar characteristics, whatever their ancestry, owing to the lack of opportunity for variety of habitat and habits. Little wonder that zoologists formerly lumped all such forms together—for instance, medusae and sea urchins, both of which are radially symmetrical.

Our example involves members of three completely distinct groups: centipedes, millipedes, and a mammal. They, too, have developed the same external form with the same biological significance in the same environment. There can be little doubt but that their ability to roll themselves into a ball provides all three with an effective means of protection, much as the torpedo shape favors the survival of fast swimmers. That all three have distinct internal structures shows clearly, however, that they must have developed independently of one another.

FIG. 42. The ability to roll up in a ball has been perfected independently by pill millipedes and pill bugs. (a) and (b) pill millipede (*Glomeris*), (c) and (d) pill bug or pill wood louse (*Armadillidium*).

More active methods of self-protection are rare among soil animals. Earthworms and snails secrete a sticky substance, and the small Symphyla millipedes eject gluey threads from two anal spinnerets. Many millipedes have a pair of defense glands on every segment, from which they can eject cyanide. It is significant that they have very few enemies.

Equilibrium in the Soil

Predatory soil animals seem to play no direct part in the transformation of vegetable litter into humus. However, they help to control the population, and in the long run no community can exist without such controls—which humans are apt to deplore, since human populations are far too often "controlled" by wars and other disasters. Among soil animals such disasters are more often useful than deleterious, and they may even have beneficial effects on other habitats. Thus, ants play an essential part in what has come to be known as biological warfare.

FIG. 43. An armadillo (*Tolypeutes conurus*) that can roll up into a ball much as do pill bugs and pill millipedes.

Soil animals are normally a fairly balanced community of meat producers and meat consumers whose density depends on the available living space (pore volume of the soil), on the primary food present in the soil (vegetable litter), and on climatic conditions governing the rate of growth, the length of life, and general "activity." Predators, parasites, fungi (as competitors for food), and diseases all help to balance the population pressure in the soil which, as everywhere in nature, is due to the continuous overproduction of offspring.

Production of Offspring

How great this overproduction really is in many species has not yet been accurately determined. Here I shall look at one instance with which I am personally familiar, that of one of the most common of soil-inhabiting springtails—*Onychiurus*. Like all the Collembola, it lays several sets of eggs during the 1-2 years of its life. Under favorable conditions a single female can produce 30-40 eggs every six weeks; in temperate regions this adds up to a minimum of 150-200 eggs a year. If these hatched out into only 50 females, all of which can lay eggs within 6-8 weeks of birth, then the offspring of a single animal could give rise to 1000-1500 mature individuals a year. If predators, parasites, and diseases did not keep a check on this explosion, the ground would quickly be bursting with springtails.

Most soil animals lay their eggs beneath stones, moss, and leaves or in cracks; earthworms and enchytraeid worms spin hardened cocoons in which the eggs are embedded (Fig. 44). Springtails scatter their egg heaps over large distances. Mites use various methods of "nursing" their young. Thus, females of the horned beetle mite *(Belba)* attach their eggs to the body of another female, which carries them until they hatch (Fig. 45). This clearly lends extra protection to the eggs, so much so

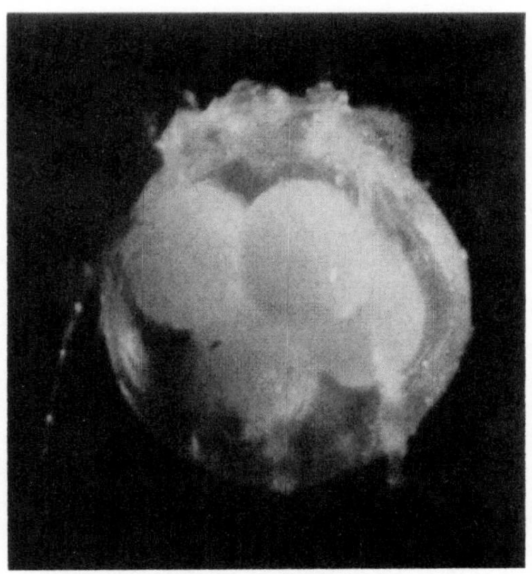

FIG. 44. Eggs in cocoon spun by a worm of the species *Enchytraeus buchholzi* (from M. Trappmann, 1950).

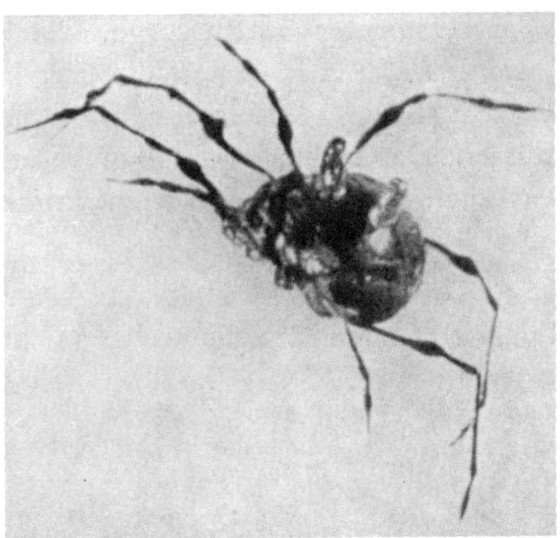

FIG. 45. Beetle mite (*Belba*) with eggs attached by another female (from F. Pauly, 1956).

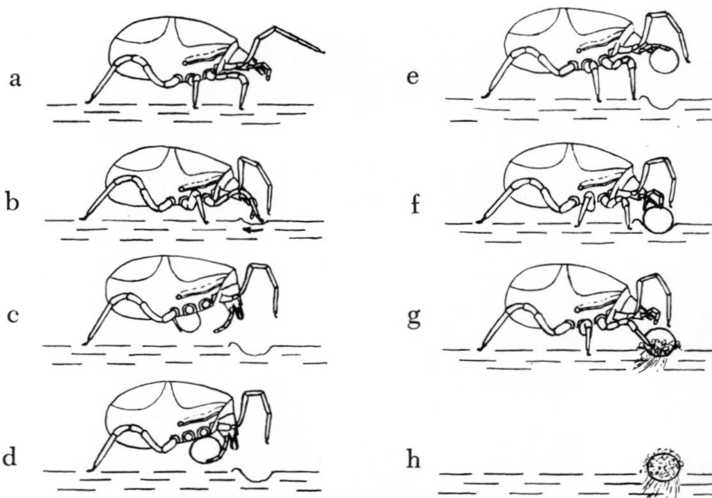

FIG. 46. Predatory beetle mite (*Parasitus coleoptratorum*) laying an egg and burying it in a groove (from A. Rapp, 1959). Actual length ca. 1.3 mm.

that this mite can afford to produce no more than a few, relatively large, eggs. Predatory beetle mites of the genus *Parasitus* have developed a different nursing instinct. They dig a small groove in the soil for every one of their eggs, position it with their legs, and then cover it carefully (Fig. 46). The final larval forms can be seen on the backs of dung beetles, which carry them from dried-up to fresh dung heaps, where they find optimum living conditions and also an abundant supply of nematodes, their favorite food.

Genuine nursing is practiced by the limbless amphibia of the order Gymnophiona (Fig. 47), by earwigs, scolopenders, and other predatory millipedes, by scorpions, japygids, and the soil bug *(Brachypelta aterrima)*. Earwig females, for instance, keep constant guard over their eggs, licking them, tending them, and transporting them

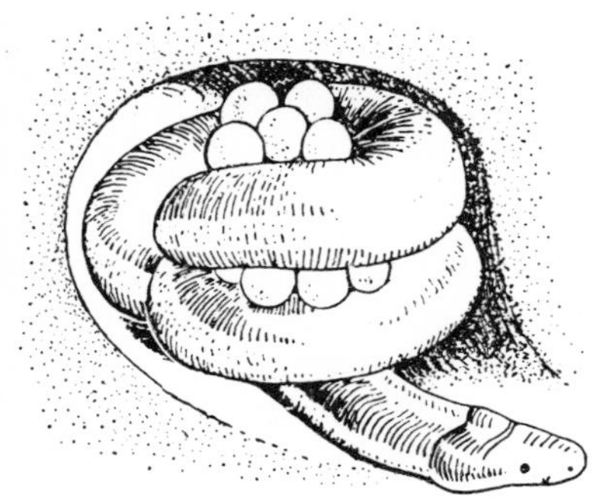

FIG. 47. Limbless amphibian (order Gymnophiona), tending eggs in its burrow.

to suitable sites (Fig. 48). All these animals produce considerably fewer eggs than those which do not tend their offspring quite so carefully (Fig. 49).

A very striking way of tending young is followed by soil bug *(Brachypelta)* females. This black bug lives in sandy ground beneath milkweeds, which provide its favorite food. Here the eggs are deposited (30-50 at a time) and guarded until they hatch out. Even after hatching the larvae are kept close together and protected by their mother, who allows the young to crawl all over

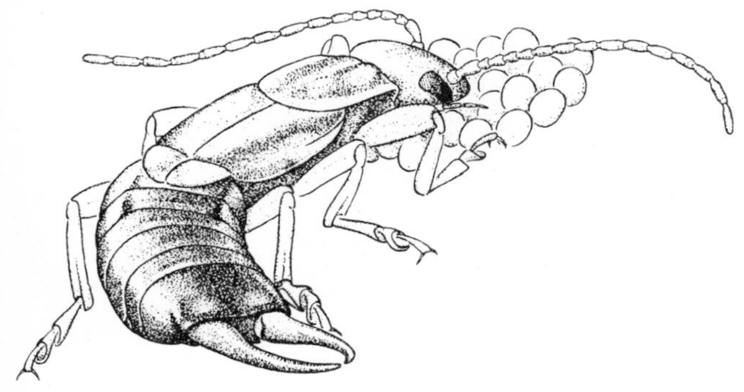

FIG. 48. Female earwig (*Forficula*) guarding and licking its eggs.

her, and from time to time secretes a droplet of fluid from her anus, which the young greedily suck up. During their first few days of free life, they take no nourishment at all, and only such larvae as mount a nursing female and suck up the droplets can survive (Fig. 50). As H. Schorr discovered in 1957, the nursing behavior of this soil bug is closely associated with a second phenomenon: the bug's intestine is inhabited by special bacteria which supplement its monotonous diet with essential proteins and vitamins. The close and mutually beneficial association of two distinct organisms is called symbiosis, and it can only continue if the "guest" can be transferred to the "host's" descendents. In the *Brachypelta* bug this is a direct consequence of the nursing method: the droplets exuded by the female contain the intestinal bacteria in a transitional form. The young thus become "infected" before they can scatter to feed on milkweed sap.

FIG. 49. Scolopender female tending young (from H. Klingel, 1960).

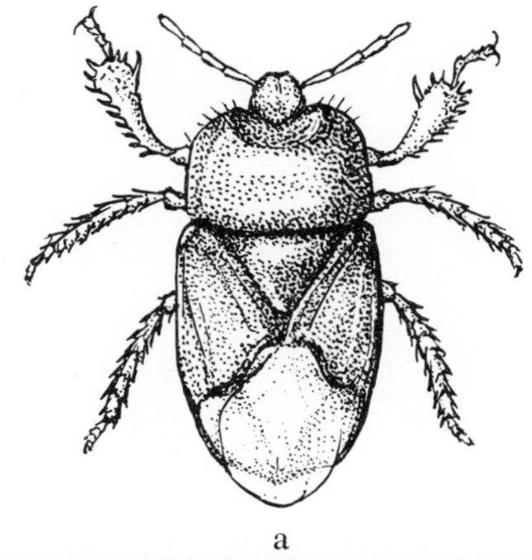

a

b

FIG. 50. (a) Soil bug (*Brachypelta aterrima*). (b) Soil bug with its young (from H. Schorr, 1957).

IV

THE MATING HABITS OF SOIL ANIMALS

In general, the males of soil animals play no part in nursing the young, and often they are rather reticent wooers, as well. This brings us to a chapter in the life of soil animals which—as elsewhere in the animal kingdom—must be counted one of the most fascinating of all.

Many soil animals mate "normally"—the sexual partners come into close physical contact and copulate. This presupposes suitable male organs capable of attaching themselves to those of the females. As a rule, the same organs also transfer the sperm into the sexual organ of the female. That is what happens in nematode worms, earthworms, enchytraeid worms, snails, higher insects, and daddy longlegs. Among these, earthworms are remarkable for two special reasons: they are hermaphrodites, and they lack special organs for the transfer of sperm. In the spring earthworms can often be seen mating beneath stones. What precisely happens is shown in Fig. 51. Each of the partners is provided with male and female organs. They attach themselves to each other by secreting a mucous envelope that seems to lace them together. The fluid is secreted from the clitellum, a glandular collar of enlarged rings about a third of the way back from the head. The worms tighten their embrace even further by boring their bristles into the partner's body.

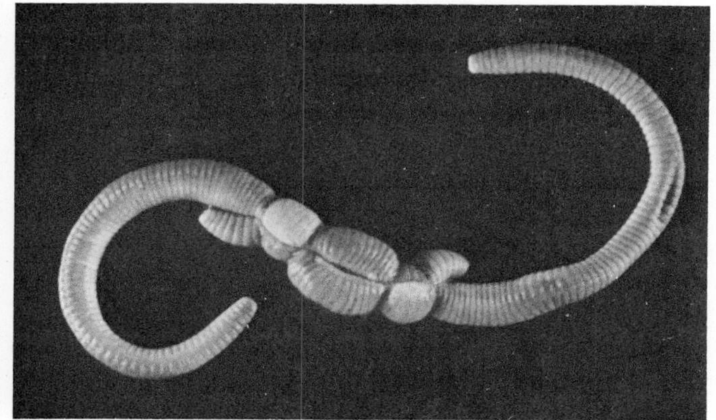

a

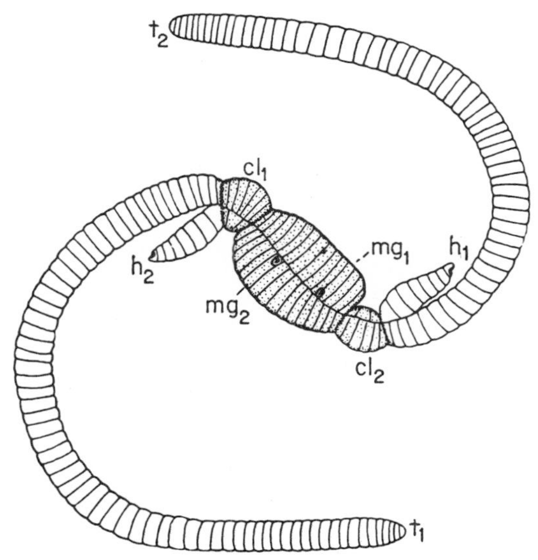

b

FIG. 51. Copulating earthworms. (a) Photographed by I. Holzapfel. (b) Drawn by Andrews: *cl* = clitellum; *h* = head; *mg* = male genital aperture; *t* = tail end. The figures refer to the individual worms. The mucous envelope is stippled.

Each earthworm has 8 sexual apertures on the ventral side: 2 male genital openings in the fifteenth segment, 2 female genital openings in the fourteenth segment, and 2 openings each of the 4 seminal receptacles (spermathecae) in the ninth and tenth segments. As Fig. 52 shows, these organs are not brought into direct contact during copulation. Instead, contractions of the ventral muscles of each partner lead to the formation of two slimy tubes from the male sexual apertures to the clitellum, that is, as far as the openings of the partner's seminal receptacles. The sperm is discharged into these tubes and carried backward by muscular contraction.

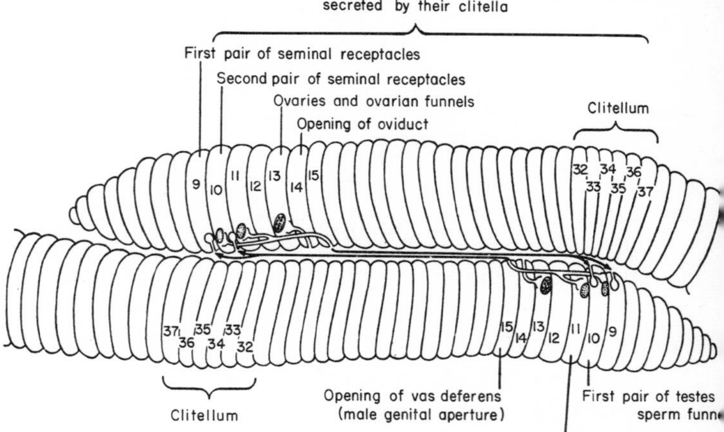

FIG. 52. Schematic view of copulating earthworms. The sexual region has been greatly magnified.

Once it has been picked up by the seminal receptacles the partners separate slowly and crawl away, each bearing the sperms of the other. Since the seminal receptacles, or spermothecae, are simply places in which the partner's sperm can be stored, copulation in earthworms does not lead directly to fertilization. The male earthworms mature several weeks before the females, that is, the sperms are formed well before the ova. This phenomenon occurs in many other hermaphrodites as well. Because of the earlier maturing of the sperms—protandry—the ova cannot be fertilized during copulation, but must be stored until they are ripe. The biological significance of this apparently unnecessary complication becomes clear when one considers that hermaphrodites are in danger of inbreeding due to self-fertilization.

During fertilization proper each earthworm secretes a ring of mucus from its clitellum and slips out of it backward. As the apertures of its oviducts and spermathecae pass the ring, the worm expels its own ova and the partner's sperm. Since the spermathecae are placed further forward, the correct sequence of events is ensured, with the result that the ova are fertilized soon after their expulsion. The ring then hardens quickly into a firm cocoon, in which the young earthworm develops.

Use of Sperm Packets

In earthworms the sperm is not simply transferred to the partner in the form of a fluid but in small globules. Such "sperm packets" are known as spermatophores, and they function widely in the animal kingdom. Generally, they are enclosed in capsules far more complicated than those produced by earthworms. More remarkable, however, is their method of transfer. This aspect of the life of soil animals has only become known during recent years. Biologists were, of course, quite aware that animal behavior is most varied in the reproductive sphere, but when they began to look more closely at soil arthropods they discovered that, though sexual behavior differs greatly from species to species, many of these animals use an identical method of sperm transfer, which I have called the "indirect transfer of spermatophores." All the males lack copulative organs, and often the means of attaching themselves to the females. Hence they cannot transfer their sperm directly, but must deposit it in packets on the ground, where the females can find and "collect" it. There are, moreover, all sorts of intermediate forms between species in which the males make some sort of "personal" contact with their partners and those in which the sexes live completely apart from each other.

In order to treat this topic systematically, we shall successively examine the reproduction characteristics of

arachnids
myriopods
Apterygota (primitive insects).

Of all the arachnids, spiders are probably those with which the layman is most familiar. Spiders only *appear* to mate in the "normal" manner. Before the male approaches the female, he first spins a small "sperm web" on which he deposits a drop of semen. He then dips his two leg-like appendages behind the mouth (pedipalps) into it and transfers the sperm to the female genital aperture in that way (Fig. 53).

Scorpions lack even this means of transferring their sperm, as they have neither a copulative organ nor specially adapted pedipalps. They normally live under stones, shunned by other soil animals, which they seize with their powerful claws or palpi and then paralyze and kill with their poisonous tail sting. At mating time the males go in search of the females and seize them by their claws. Then they take their partners for a "walk" that may last several hours—and that the famous French entomologist Jean Henri Fabre has called the *promenade à deux*. During it, the male keeps pulling the female to him and stroking her abdomen with his front legs. Less romantically, he also keeps stinging her claw joint, for

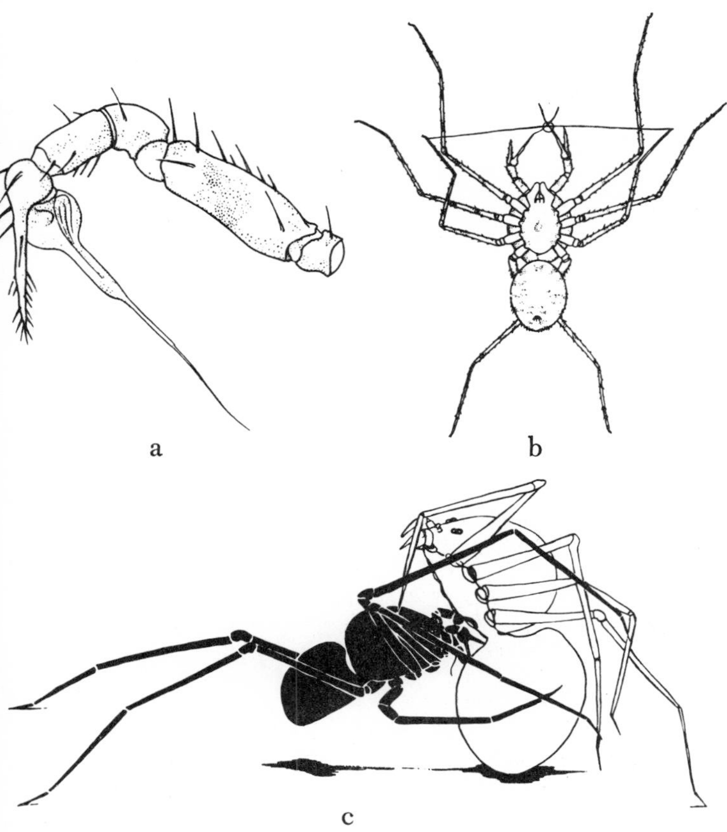

FIG. 53. Male spider (*Scytodes*). (a) Pedipalp with special appendage for transferring sperm. (b) Filling its pedipalps with sperm, which it has previously deposited on a special "sperm web" (from S. Dabelow, 1958). (c) Placing its pedipalps in the genital aperture of the female (from S. Dabelow, 1958).

a

b

FIG. 54. Mating of scorpions. (a) The male (left) stroking the female's abdomen. (b) Stinging her right claw (from H. Angermann, 1957).

reasons that still elude us (Fig. 54). The female, however, does not seem to react to this affront. Finally, the male presses his abdomen against the ground and then lifts it up slowly: in the process his genital aperture has expelled a small rodlike structure, thickened on top and provided with a prominent "wing." Once he has stuck this rod to the ground, the male moves slowly backward pulling his partner along until she touches the rod with her abdomen (Fig. 55). Upon doing so, she suddenly jerks backward, while the male releases her and runs away. The whole chain of actions becomes comprehensible once we realize that the thickened end of the rod is a kind of seminal receptacle and that the wing is a kind of lever for opening it. As the female presses on it with her abdomen, the lever tears open a thin wall, and two sperm globules ooze out from the receptacle into the female sexual aperture.

Here we have our first instance of indirect spermatophore transfer: the male places a sperm packet on the ground and the female then collects it. Scorpions produce unusually large spermatophores with an unusually complex structure and transfer mechanism. In a small Mediterranean scorpion *(Euscorpius italicus)* the rods are 5-6 mm high. Close examination has shown that they consist of two halves. Their relatively large size makes it unlikely that the males would have many of them available at a time. As far as is known, they are ready for mating at least once every four days (Figs. 56, 57).

In the male reproductive organs, in addition to germinal glands, scorpions have a number of glandular spaces representing precise "molds" of the spermatophores. In them the two halves and the sperm packets arise separately. The formative period is at least 3 days. Only when the rods have been expelled from the male's body are the two halves cemented together with a sticky substance and filled with sperm.

Indirect spermatophore transfer in whip scorpions was

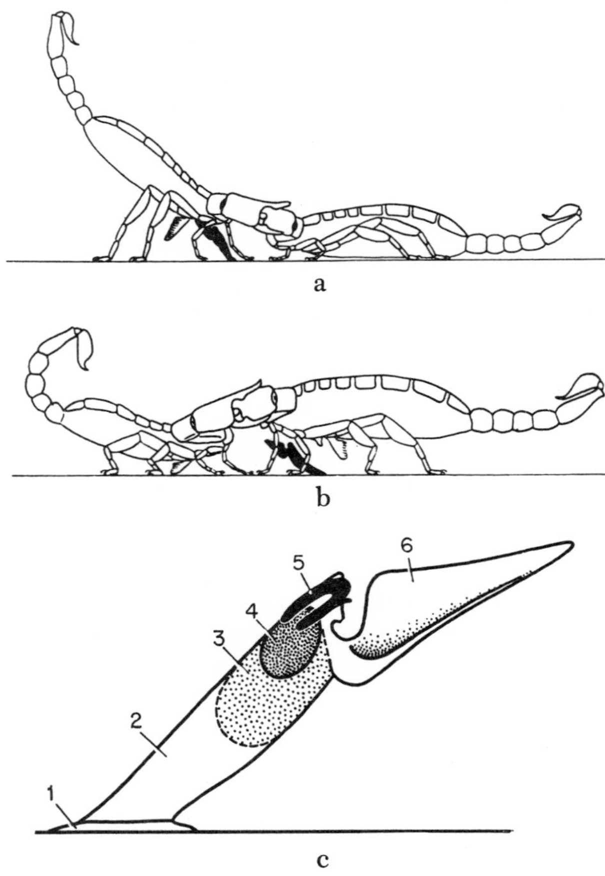

FIG. 55. Mating of scorpions. (a) Male having deposited a spermatophore (black). (b) Male pulling the female across the spermatophore. (c) Diagramatic view of spermatophore: 1 = basal plate; 2 = rod; 3 = seminal receptacle; 4 = sperm globule; 5 = opening mechanism; 6 = wing-shaped lever (from H. Angermann, 1957).

first observed as recently as 1958 by H. Sturm in Colombia. The scorpion concerned *(Trithyreus sturmi)* is a typical soil inhabitant of the tropics. Taxonomists had noticed that males of the genus *Trithyreus* have a strange

FIG. 56. Spermatophore of Mediterranean scorpion (*Euscorpius italicus*) unopened. Actual length ca. 5 mm (from H. Angermann, 1957).

FIG. 57. Scorpion spermatophore (*Euscorpius italicus*) opened by pressure on wing-shaped appendage and revealing seminal receptacle and sperm ball (from H. Angermann, 1957).

"button" in the place of the usual anal appendage. In these arachnids the male pursues the female and belabors her rear with his two pedipalps. Then the female turns round and pursues the male in her turn. In so doing, she

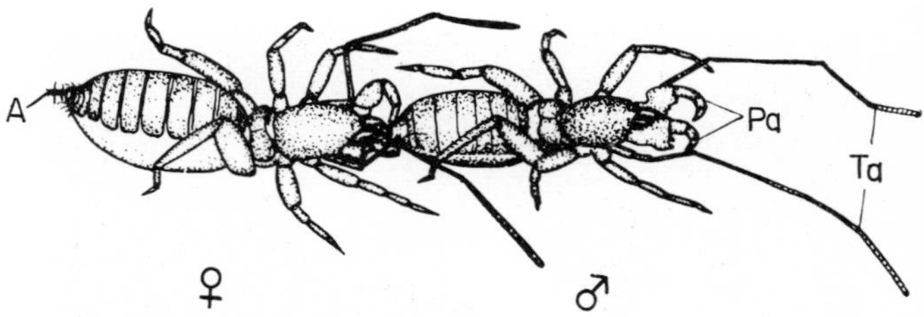

FIG. 58. Male and female whip scorpions (*Trithyreus sturmi*). The female has fastened herself to the male's tail button. A = anal appendage of female; Pa = claw-shaped palps of male; Ta = tactile legs of male. Actual size ca. 5 mm (from H. Sturm, 1958).

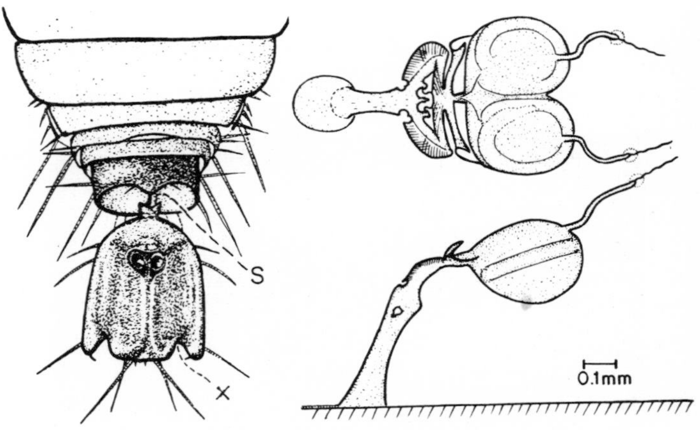

FIG. 59. Head appendage of male whip scorpion (*Trithyreus*). The female digs her claws in at S.

FIG. 60. Spermatophore of whip scorpion from the top and the side. Note the two sperm globules.

digs her jaws (chelicerae) into two grooves in the male's "button." If we remember what happens with scorpions, it is easy to foresee what happens next: the male deposits a rod-shaped spermatophore and pulls the female across it, whereupon she picks up two sperm packets in her vulva. Here, however, there is no lever to open the spermatophore (Figs. 58-60).

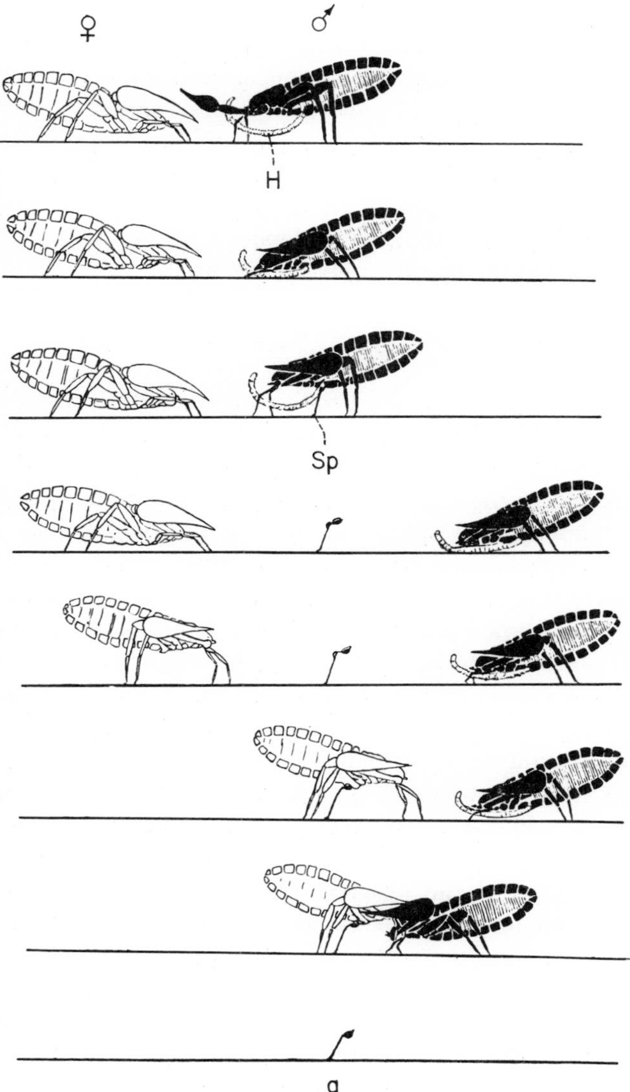

FIG. 61. Mating in a pseudoscorpion (*Chelifer*); H = hook-shaped stimulating organ of male; Sp = spermatophore; a = rod of spermatophore left behind after mating dance (from Vachon, 1938).

The behavior of pseudoscorpions resembles that of scorpions rather than of whip scorpions: the male seizes the female, performs a long dance, deposits a rod-shaped spermatophore, and pulls his partner across it. Some species omit all physical contact; the male simple emits two hook-shaped tubes and dances in front of his partner. Attracted by olfactory stimuli, she soon begins to follow his movements at some distance, and in that way is tempted to the spermatophore (Fig. 61).

Mating Without "Personal" Contact

Scorpions, whip scorpions, and pseudoscorpions are closely related and form the first order in the class of Arachnida. Mites, to which we now turn, make up the tail end of the arachnid section. So large is the number of families and species in this order, that it has not been fully surveyed to this day. Most species are carnivorous and reproduce their kind by direct sperm transfer; the horned beetle mite *(Belba)*, however, has a form of indirect sperm transfer normally restricted to springtails. Here the fertilization of the ova does not involve any kind of "personal" contact. Even after long periods of isolation in the laboratory—which were necessary at the start of the investigations since the partners could not be told apart until sex was determined by observing selected groups of two—the animals take no notice of each other. If the observer is lucky, however, he may catch one of them moving about in a most peculiar way. The mite presses its short and squat body to the ground, raises it again slowly, remains in the normal attitude for a brief time, then rears up and finally walks off. If we look at the places it has just visited under a powerful microscope, we discover small, shiny droplets suspended on thin stalks (Figs. 62-64). These are typical spermatophores of horned beetle mites. What is remarkable about them is that they are deposited by the male no matter

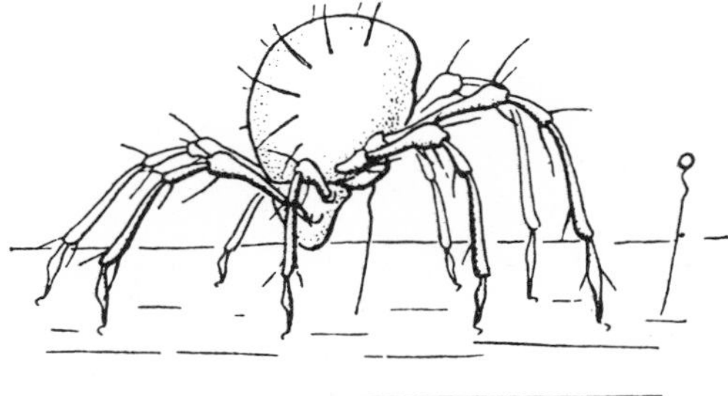

FIG. 62. Male beetle mite (*Belba geniculosa*) depositing its "sperm stalks" (from F. Pauly, 1956).

whether a female is present or not–indeed, even males kept in strict isolation for a long time continue to produce such spermatophores. Nor need the female be "shown" the spermatophores by a male. She finds them by herself as she carefully walks about and collects what droplets she comes across by brushing against them with her wide-open genital aperture.

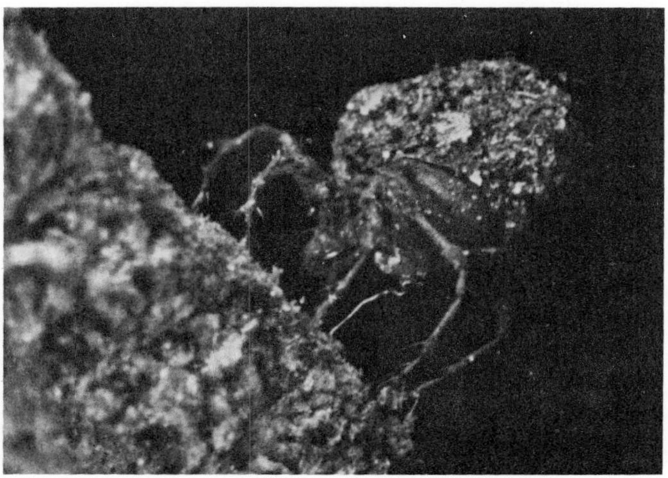

FIG. 63. Photograph of male beetle mite (*Belba geniculosa*) with "sperm stalks" (F. Pauly, 1956).

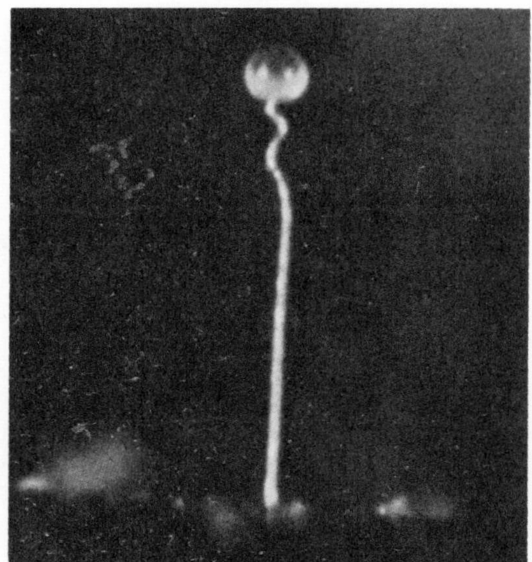

FIG. 64. Spermatophore of beetle mite (*Belba*); actual length of stalk: 0.8 mm.

The droplet may remain suspended for quite some time before it is discovered by a female. But this can only happen in the soil, where constant conditions of high humidity prevent dessication—one reason why indirect spermatophore transfer is considered a biological characteristic of soil animals.

Since these beetle mite *(Belba)* males deposit their spermatophores at random, the collection of them by females depends on the presence of large numbers of them along her normal path. A single male can produce a good dozen or more during a day. No wonder that at times—especially in spring and late autumn—the soil is thickly covered with such structures. F. Pauly deserves great credit for having first recognized one of these spermatophores. Other observers probably mistook them

FIG. 65. One of the smallest diplopods (*Polyxenus lagurus*), whose striking bristles make it one of the most attractive of all millipedes (from K. H. Schömann, 1956).

for conidia—the reproductive bodies of certain fungi, which they greatly resemble.

Myriopods: We must distinguish between herbivorous and carnivorous myriopods, which differ considerably in morphological respect—in the present context the most important distinction is that the carnivores have their genital apertures in the front part of the body, the herbivores in the rear.

In the large vegetarian diplopods—such as julids and glomerids—mating involves close physical contact. Nevertheless, the sperm is not transferred directly from genital opening to genital opening, but by means of specially adapted legs—the so-called gonopodia.

The smallest and most primitive diplopods (order Pselaphognatha) use a method of sperm transfer that is

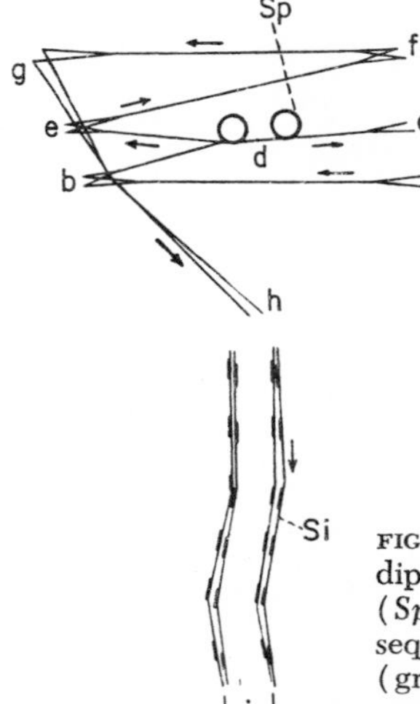

FIG. 66. The web of a *Polyxenus* male diplopod with two droplets of sperm (*Sp*). The arrows and letters indicate the sequence of events. *Si* = signal strip (greatly reduced) (from K. H. Schömann, 1956).

unique in the entire animal kingdom. These animals live underneath the bark of trees and in the loose, dry humus inhabited by algae. A relatively unknown species found in Central Europe *(Polyxenus lagurus)* is one of the most beautiful of all millipedes.

The sexual behavior of these animals has become known only recently, partly because the European species produce offspring parthenogenetically (by "virgin birth"). Hence European species generally consist of females only. But in 1954 K. H. Schömann accidentally came across a mating pair on the island of Sylt and was able to study their sexual behavior.

Males and females alike have two genital apertures on special papillae at the base of their second pair of legs. Morphologists were convinced that the more pointed sexual papillae of the males were penes—normal organs of copulation—but Schömann's observations suggest that they have a quite different function and that males never have direct contact with their "partners."

The male searches for a small depression in the ground, and when he has found one he begins to move about in a strange manner. He pushes the front of his body into the depression several times, turns round, and walks away in a straight line. If we look at him from the side, we discover that he presses his "penis" into the right and left walls of the depression, and then spins a web across. Once he has turned round, he pulls out two thick strips of mucus from special pockets in his eighth and ninth pairs of legs to produce a straight "road" some 1.5 cm in length (Figs. 65, 66).

On the web itself are two glistening droplets, whose nature becomes clear as soon as a "mature" female happens to pass by. The moment her antennae touch one of the broad strips she becomes agitated and starts feeling her way to the web. The web itself is clearly a stop signal, for the moment she reaches it she stops running and initiates search movements with her protruding genital papillae (vulvae). The search culminates in the discovery and collection of the two droplets of sperm (Fig. 67). In principle, therefore, the whole process is just another instance of indirect spermatophore transfer. The precise function of the strips is shown by the fact that females will ignore the sperm droplets in the absence of such signals—indeed, they will often "stumble" over the droplets without taking the least notice.

Another oddity is the behavior of the males to their own spermatophores. They, too, can be observed following along the strips, only to eat up what sperms they discover. Then they immediately spin a new web and deposit fresh droplets of sperm. In the process, they often produce thick "roads" made up of a host of strips (Fig. 68). The biological significance of their behavior is obvious: it increases the signal strength of the strips, and it provides an ever-fresh store of sperm for the females.

Stranger still is the "love life" of another small group of

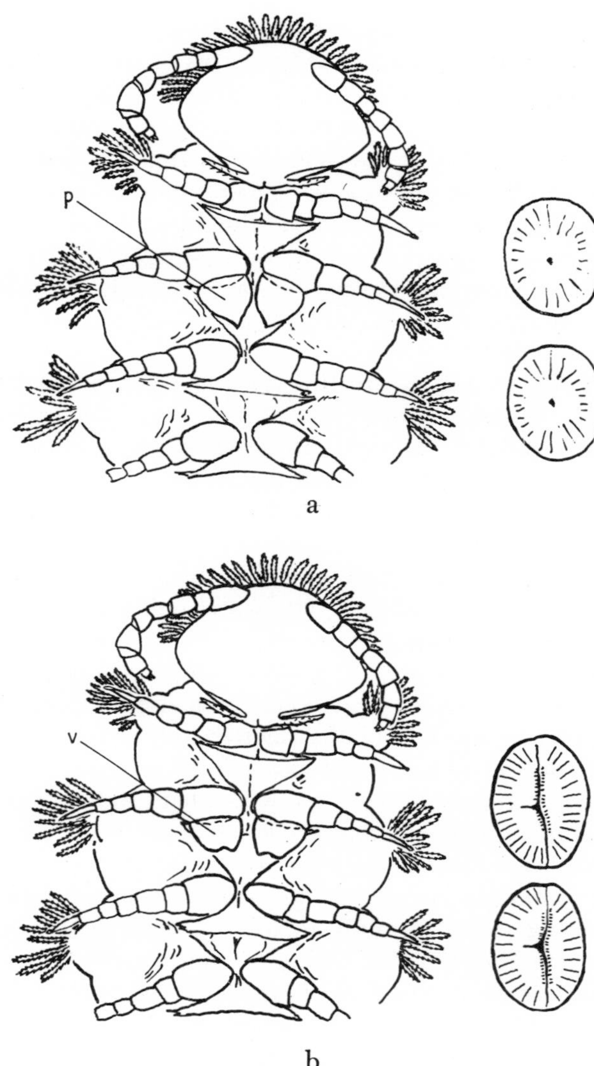

FIG. 67. Male (*p*) and female (*v*) genital appendages of *Polyxenus* diplopods at the base of the second pair of legs. (a) Male, with top view of the tips of the genital papillae (right). (b) Female, with top view of the sexual apertures (right) (from K. H. Schömann, 1956).

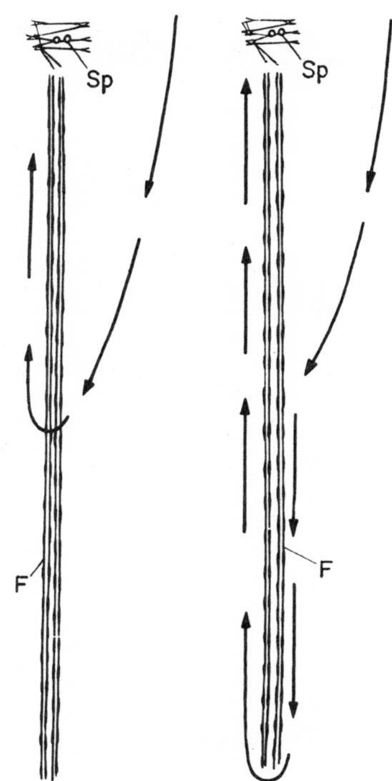

FIG. 68. Strips (F) produced by the *Polyxenus* male. The arrows show how females are guided to the sperm (*Sp*) droplets (from K. H. Schömann, 1956).

myriopods, the Symphyla, probably the most "genuine" soil dwellers of the whole class. They are small (maximum length 1 cm), devoid of pigment, and blind, and they generally inhabit loose humus. They like warm climates and are rare in temperate zones.

I made a special study of their sexual biology in an effort to prove that they, too, practice indirect spermatophore transfer. I failed in this, possibly because I, too, was working with parthenogenetic females. More recently, however, Mme L. Juberthie-Jupeau (1956-59) succeeded in discovering the sexual behavior of the Symphyla. The male of this myriopod produces a thin stalk from the anterior sexual aperture (Fig. 69) and attaches

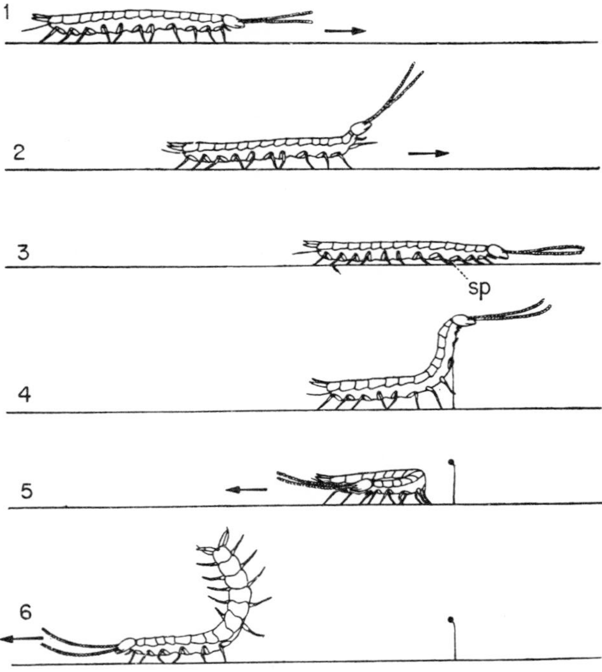

FIG. 69. Male of a small blind myriopod (*Scutigerella*) depositing its sperm stalks (from L. Juberthie-Jupeau, 1959).

a seminal droplet to it. When the female comes across it, she does something altogether unexpected: she devours the droplet. She does not swallow it completely, however, but deposits it in a special cheek pouch. The significance of this act becomes clear when she subsequently produces her ova: each ovum is removed individually from the vulva, placed in her mouth, and then attached to any protuberance in the ground (Fig. 70). Mme Juberthie-Jupeau was able to show that the ova are fertilized by having a small portion of sperm smeared over them.

Despite this remarkable variation the Symphyla myriopods, too, produce stalked spermatophores and hence

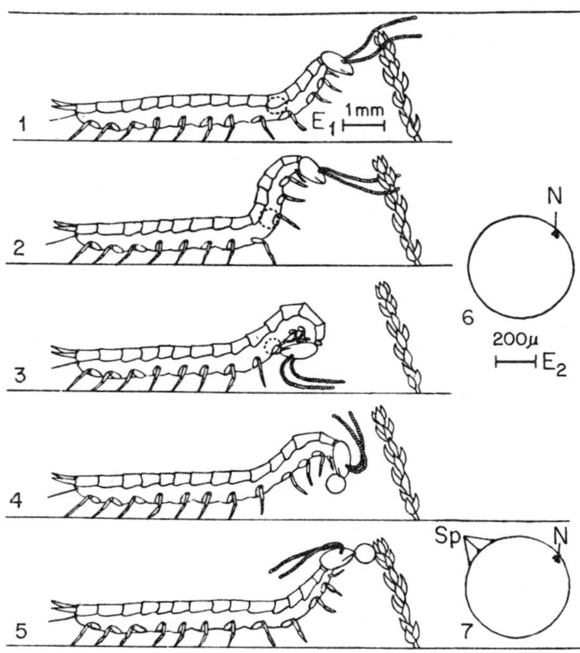

FIG. 70. Female blind myriopod (*Scutigerella*), having "eaten" a spermatophore and deposited the sperm in a special cheek pouch, fertilizes each ovum by smearing sperm over it (from L. Juberthie-Jupeau, 1959). (1) and (2) The female looks for a suitable plant stalk; (3) she removes the ovum from her sexual aperture in the third segment of the body; (4) and (5) she sticks the ovum to the plant; (6) and (7) ovum greatly magnified. *Sp* = sperms on the surface of the ovum.

practice indirect spermatophore transfer. As we shall see, their case is particularly important in phylogenetic terms.

Indirect spermatophore transfer is also the rule among predatory myriopods (Chilopoda), as H. Klingels has shown. The typical soil dwellers among them are the Geophilidae, most of which live in deeper layers of the

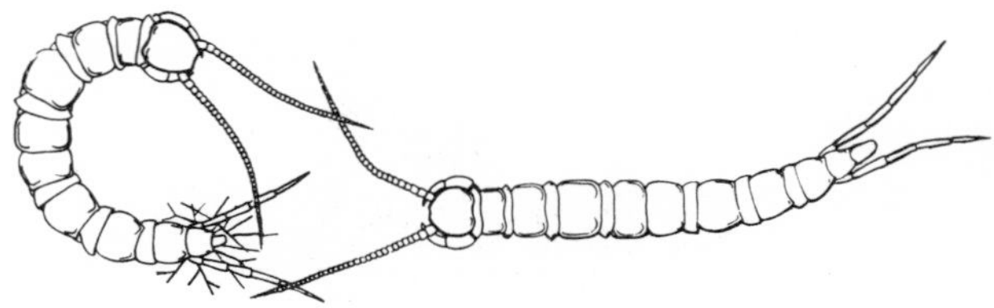

FIG. 71. Centipedes mating (*Lithobius*)—legs are omitted for greater clarity. The male is turning round, thus signaling the female to move forward. Note web beneath tail of male.

soil, where they like to prey on soft-skinned larvae and worms. Their bodies are specially adapted to their habitat—their threadlike shape enables them to slip through the narrowest of cracks (cf. Fig. 12a).

Males and females meet in narrow passages. The male produces a small web with its posterior "penis" and deposits a spermatophore on it, while the female waits patiently nearby. Once the male has performed these acts he moves away; the female then pursues him across the web and collects the droplet of sperm in her sexual aperture. Since the whole process takes place in a narrow passage, the female can hardly miss the sperm.

Centipedes of the genus *Lithobius,* which are common in loose soil, leaf litter, and underneath stones, mate in a similar way. The male and female form a circle and strum on the partner's rear end with their antennae. The male will occasionally whip up its tail or flap it to and fro. After about an hour the pair starts a kind of lover's walk: he runs off a short distance, waits for her to follow and to thump him again, and then runs off again, and so on. Finally, he stops in his tracks, spins an asymmetrical web with his "penis," and deposits an encapsulated spermatophore upon it. Then he crawls back several millimeters and produces several broad strips of mucus. Next he inclines his head and thorax backward toward the waiting partner and touches her antennae with his own

(Fig. 71). This is obviously the signal for her to advance. She crawls over his tail, whereupon he advances another few millimeters, releases a spermatophore, which the partner seizes with her gonopods and tears from the web (Fig. 72). The whole process may last three hours or more, and the most striking thing about it is the way in which the male signals his partner to advance at the right moment (Fig. 73).

Closely related to the *Lithobius* centipede is the more southerly scolopender. These centipedes also live beneath stones and in self-made passages, and they, too, transfer spermatophores indirectly. The male and female form a

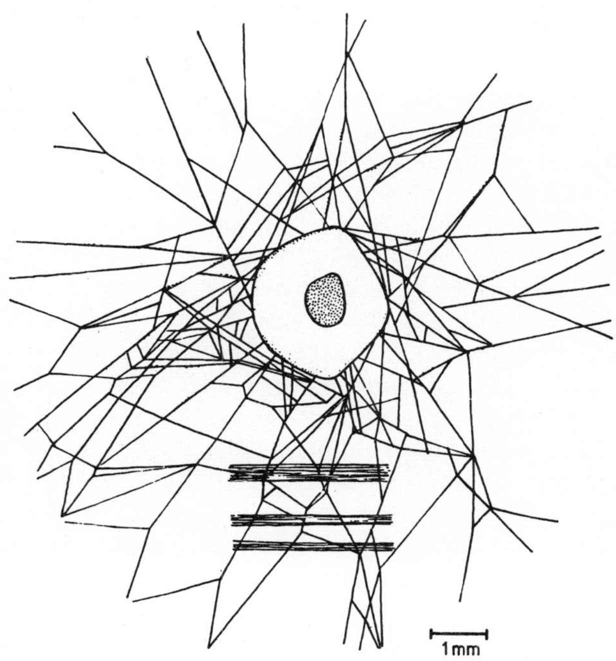

FIG. 72. Web and spermatophore of centipede (*Lithobius*). Note the 3 cross strips which are a stopping signal for the female (from H. Klingel, 1959).

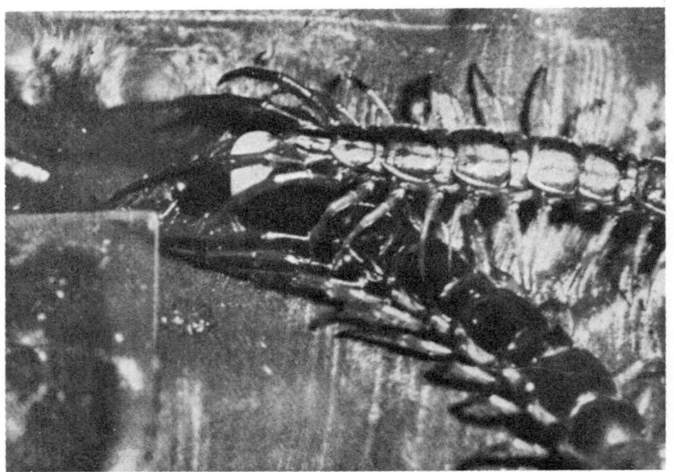

a

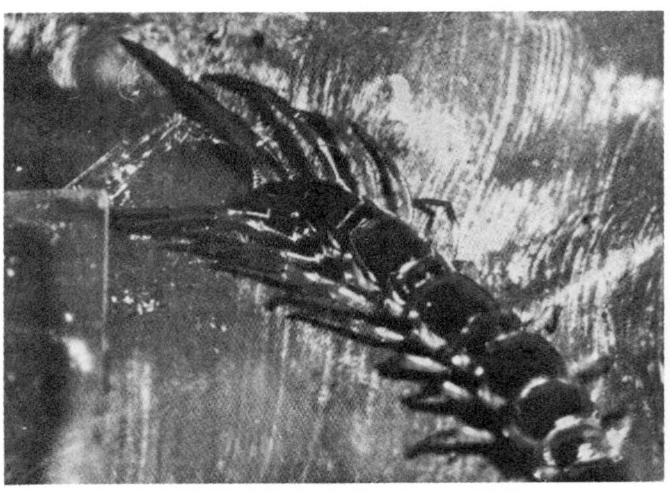

b

FIG. 73. Photographed mating of centipede (*Lithobius*). (a) Female crawling over partner's tail and removing spermatophore from the web. (b) The female has removed the spermatophore and gone away, leaving the male behind. The web can still be seen between his hind legs, as can the hole in the place where the spermatophore had been (from H. Klingel, 1959).

circle and touch tails (Fig. 74). If the meeting place is too confined, each offers its tail to the partner. The male then spins a web round the walls of the passage (Fig. 75) with its "penis" and deposits a solid, bean-shaped spermatophore upon it. When he slowly crawls away, his partner follows him across the web. The moment she touches it with her hind legs, she comes to a stop and begins to search for the spermatophore with her extended vulva (Fig. 76). At the slightest contact the spermatophore bursts open and empties the sperm into the female genital organ.

Though all observations so far indicate that chilopod reproduction involves mating, the partners have no special organs for clinging to each other. Instead, they

FIG. 74. Flashlight photograph of "courting" scolopender centipedes through a glass plate (from H. Klingel, 1960).

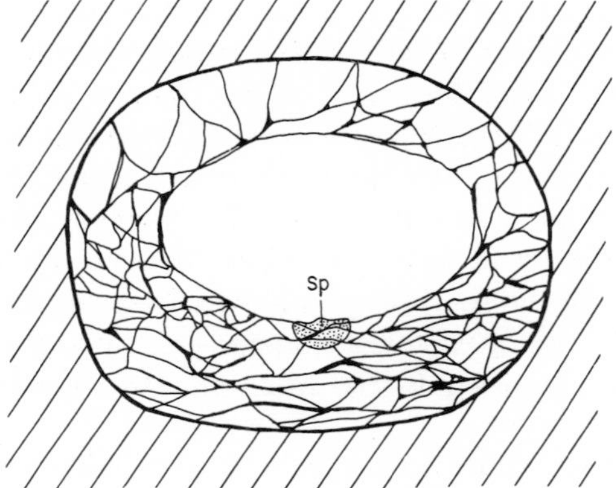

FIG. 75. Cross section of scolopender burrow with web spun by male; *Sp* = spermatophore (from H. Klingel, 1960).

carry on intensive love play with their feelers and produce webs that lead the female to the spermatophore.

Primitive insects (Apterygota) all lack wings and are mostly typical soil dwellers. Here, ecological uniformity goes hand in hand with sexual uniformity: all species investigated so far transfer their spermatophores indirectly. We shall begin by looking at the Thysanura or bristletails. Taxonomically speaking, they are the highest order of this primitive division. They live between stones and in leaf litter and feed on algae and plant matter. The group includes the machilids and the silverfish (lepismatids), which are common indoors and capable of doing serious damage to books, by eating the glue in their bindings.

The peculiar love play of machilids—a cosmopolitan family of primitive insects—has been investigated by H. Sturm (1955). Male and female begin by touching each other with their feelers, the male being more active at this stage. He is obviously concerned to discover whether his partner is in the right mood since, as we shall see, only in that event are his efforts likely to be crowned with success(Fig. 77). The female partner must not only remain in his immediate vicinity, but take up a fixed position as he steps back a few paces and fixes a thread to the ground with his "penis." With his abdomen raised he then advances again, while stretching the thread on which 3-4 small sperm droplets can now be seen. He must "persuade" his partner to collect the droplets from the thread he is holding out to her. This he does by intense tapping of his feelers, palps, and front legs, as a result of which she is driven first into a semicircle and finally into a position parallel to the thread he is holding up to her all the while (Fig. 78). She then feels the thread with her anal appendages, opens her ovipositor, and searches for the sperm, which she generally manages to collect (Figs. 79, 80).

The sexual life of lepismatids—soil-dwelling silverfish—has not yet been fully investigated. However, in the species *Lepisma saccharina*, love play begins very much as it does in machilids. Both partners feel each other all over, with the male being the more active partner of the two. He then runs past the female, slaps his tail against her head several times, and quickly withdraws (Figs. 81, 82). If she is receptive she soon moves toward him. The more difficult part of the program follows. The male

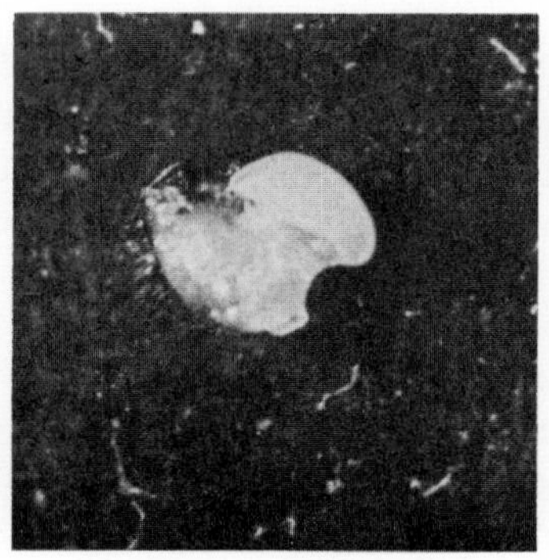

FIG. 76. Spermatophore of scolopender. Actual size 3-4 mm (from H. Klingel, unpublished).

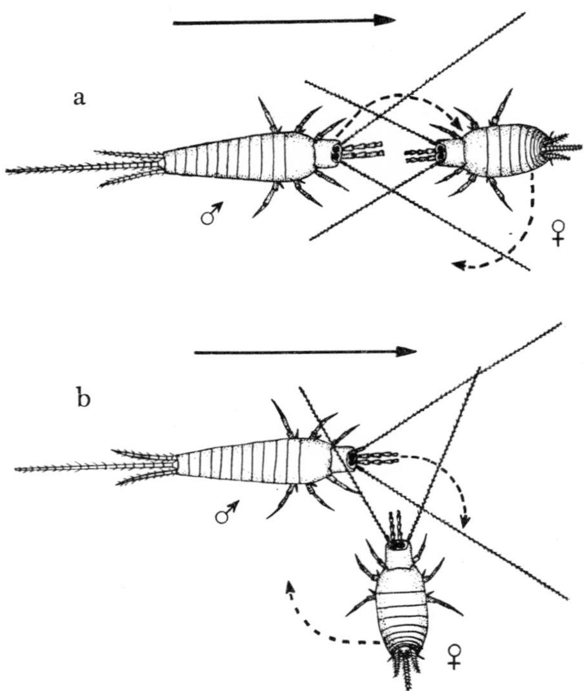

FIG. 77. (a) Foreplay among machilids. The male keeps advancing until he succeeds in swinging his partner round. (b) The female keeps touching his antennae and palps with her own (from H. Sturm, 1955).

looks for a small wall or rise, and having found one he once again runs past her, turns round, and beats his tail against the wall several times. As he does so he pulls out several threads, stretching them from the wall to the ground (Figs. 83, 84). Unlike the machilid, he does not keep the threads taut by himself. Under the threads he deposits a spermatophore. The whole operation has to be performed at great speed, for his partner has followed him and now that he no longer blocks her path she rushes past with raised tail. In so doing she touches the threads, comes to a sudden halt, lowers her abdomen, and begins to search for the spermatophore. Generally, she succeeds in her task. If the threads are torn experimentally, she will race past the spermatophores, which proves clearly that the threads are a kind of signal to stop.

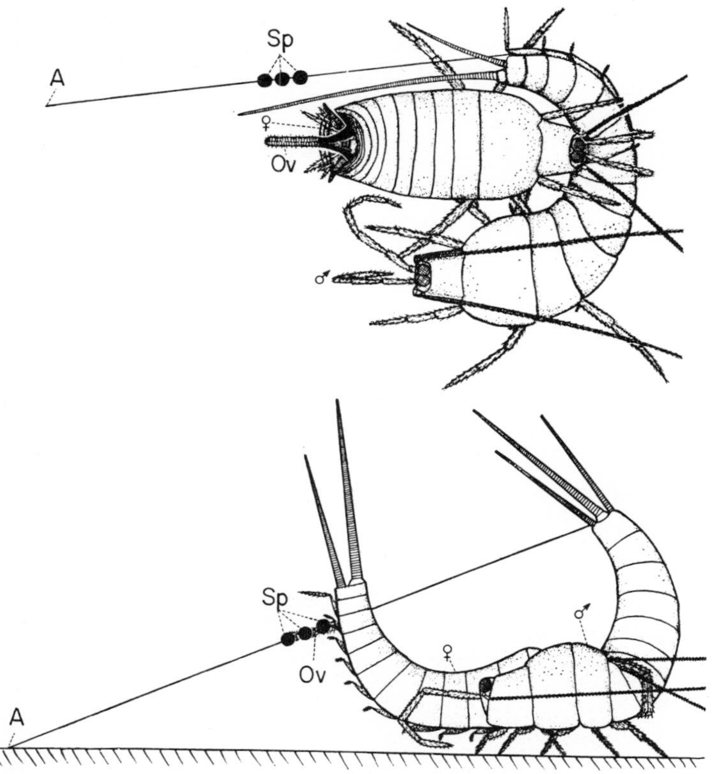

FIG. 78. During foreplay the male machilid has attached a thread to the ground with his "penis," pulled it taut at an angle, and deposited several drops of sperm (*Sp*) on it. With his palps and thorax he pushes his partner round until her ovipositor (*Ov*) comes in contact with the sperm. Top: the pair in their final position seen from above; below: side view (from H. Sturm, 1955).

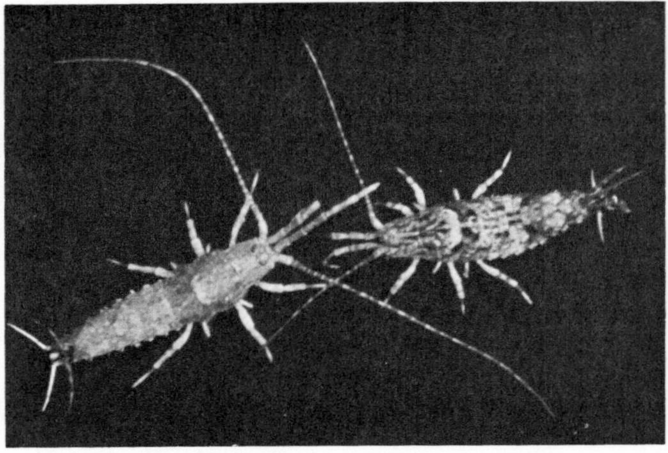

FIG. 79. A phase in the foreplay of machilids—the male at the left (from H. Sturm, 1955).

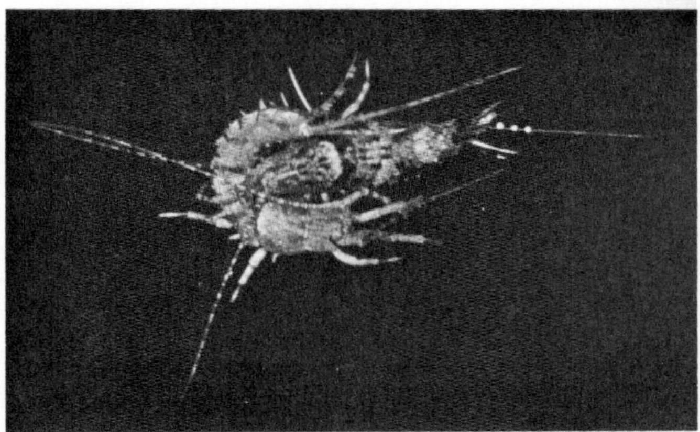

FIG. 80. Final position adopted by mating machilids. As in Fig. 79, the animals have been photographed on black velvet, which shows up the threads and sperm droplets (from H. Sturm, 1955).

The sexual behavior of the second order of the Apterygota—the springtails or Collembola—has been studied far more closely. Springtails are not only one of the most important groups of soil inhabitants, but also one in which the phenomenon of indirect spermatophore transfer occurs in all its varieties.

The indirect spermatophore transfers discussed so far fall roughly into the following categories:

1) The partners form a pair by
   a) grasping each other tightly (scorpions, whip scorpions, some pseudoscorpions) or
   b) by "communicating" with each other through protracted love play (scolopenders, centipedes (*Lithobius*), geophilids, some pseudoscorpions, machilids, silverfish).

2) The partners take no special notice of each other and act independently during both the production and the collection of the sperm (beetle mites, Pselaphognatha, Symphyla).

Almost every one of these methods is found in Collembola, which is not surprising as they form a particularly varied group of soil-living animals. There is hardly a terrestrial habitat from which they are absent. Some of them have even taken to the surface of the water.

With one of these—*Sminthurides aquaticus*—we begin our discussion. This small springtail, which looks like a yellow dot, moves about on the surface of ponds and pools covered with duckweed and disappears with a mighty jump at the slightest disturbance.

The males are considerably smaller than the females. Unlike most other springtails they can be readily identified by their antennae, which have been modified into clasping organs (Fig 85). During the summer they are commonly in pairs, the male clutching the female with his antennae. While the female rushes about the duckweed, the male seems to be floating in the air or lying with his back on the female. Nevertheless, after a time, the male takes charge and starts pulling his gigantic mate to and fro in what looks like a ritual dance. If she obeys, he immediately ejects a stalked spermatophore and leads his partner to it by tugging at her antennae or by swinging her round through an arc of 180°. Though the male seems

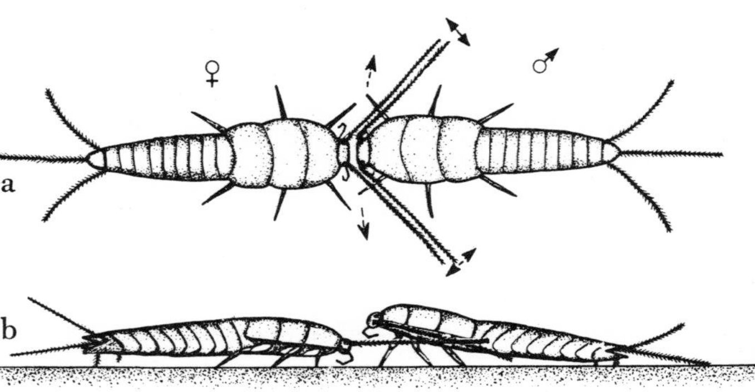

FIG. 81. Silverfish during foreplay, touching heads and vibrating their antennae (see arrows). (a) Top view; (b) side view. Actual size *ca.* 1.2 cm (from H. Sturm, 1956).

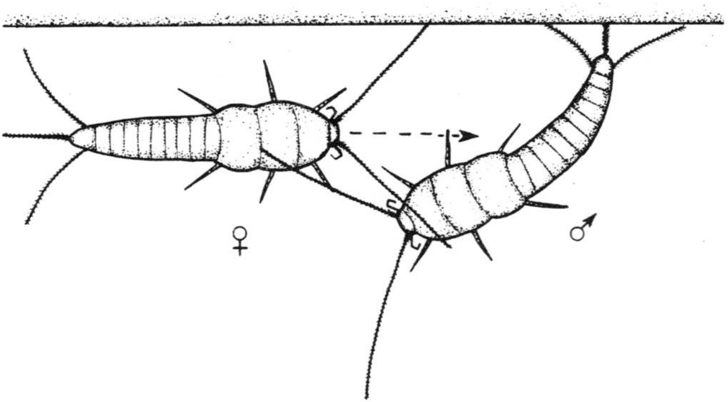

FIG. 82. In the final phase the male silverfish blocks his partner's path by beating his tail several times against the wall (top). In this process he produces several threads and deposits a spermatophore on the ground (from H. Sturm, 1956).

to be exerting himself tremendously, the whole operation usually proves successful; the female is brought to the spermatophore and collects it in her vulva (Fig. 86). Only when the male swings the female is there an oc-

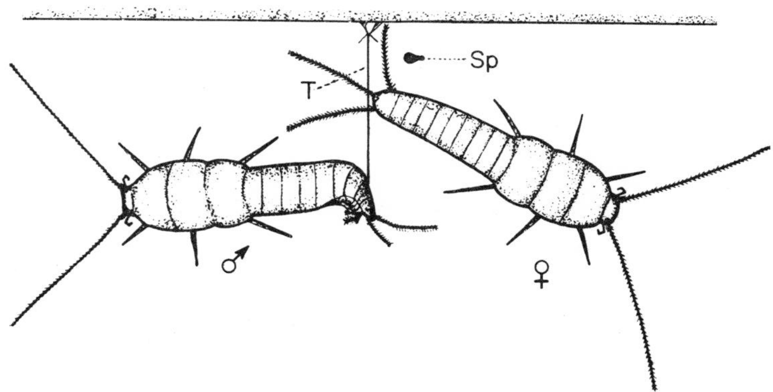

FIG. 83. The female silverfish advances until her raised tail meets one of the threads (*T*), when she is brought to a sudden halt. She immediately starts searching for a spermatophore (*Sp*) with her genital aperture.

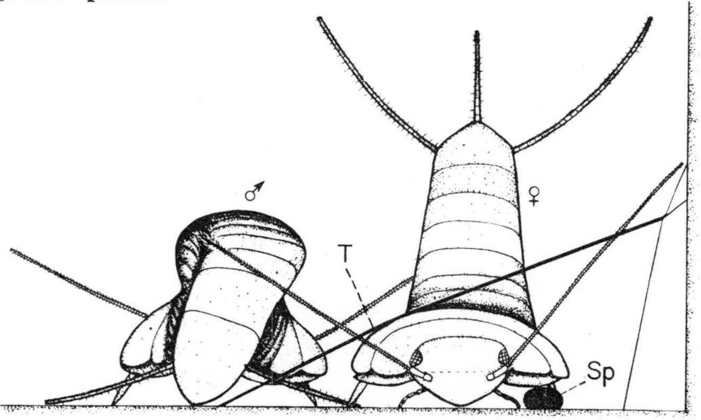

FIG. 84. Bottom view of mating silverfishes. The female (right) is trying to pass the male but her abdomen is stopped by the thread (*T*). The spermatophore (*Sp*) can be seen on the ground beneath her. The thread, which runs from the right wall to the ground, is not held by the male (from H. Sturm, 1956).

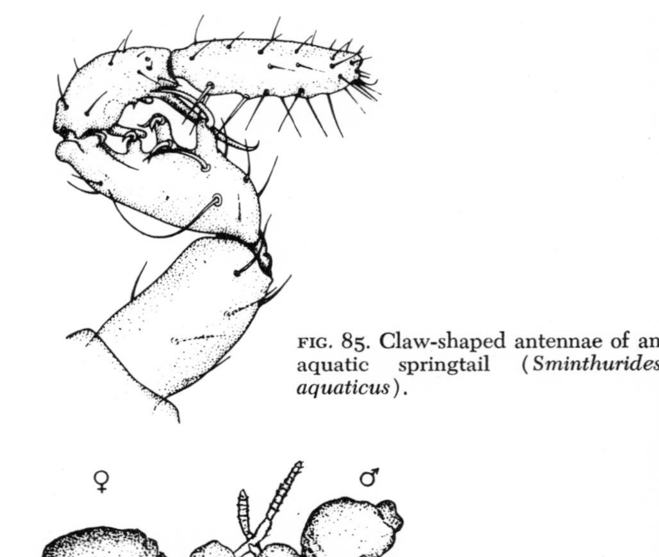

FIG. 85. Claw-shaped antennae of an aquatic springtail (*Sminthurides aquaticus*).

FIG. 86. Springtail mating (*Sminthurides aquaticus*) (from E. Handschin).

casional failure, particularly when the difference between the partners is so great that, as the couple swing round, the female's vulva lands far beyond the spermatophore (Fig. 87).

Though this springtail (*Sminthurides aquaticus*) lives on the surface of the water, there are good ecological reasons for mentioning it here. Its aquatic habit is a secondary adaptation—all its near relatives are terrestrial. Moreover, many of these terrestrial relatives can take to the water for longer or shorter periods, for they have a moisture-resistant skin and are too light to break the surface tension of the water. Indeed, as soil dwellers, they are often immersed in water after downpours, which their survive inside a small air bladder.

In respect to its sexual life, however, this aquatic form is unique among springtails—the only other species of springtail to grasp its mate by means of the claw-shaped antennae is *Dicyrtomina minuta.* Its mating behavior (discovered by H. Mayer, 1957) is of particular theoretical importance.

In autumn this springtail often appears in deciduous and mixed forests in great numbers, so that leaf litter and tree trunks seem to be completely covered by them. If we watch the behavior of these animals, each the size of a pinhead, more closely, we discover the existence of peculiar "family relationships." The females search for their food, apparently quite indifferent to their partners, rest at intervals, and expend a good deal of energy on self-grooming. The males are far more restless, rushing from leaf to leaf or from tree trunk to tree trunk in a frantic search for partners. They make a beeline for every moving object the size of a pinhead and touch it with their feelers. If the object should turn out to be the head of a real pin—which does not happen unless a

FIG. 87. Springtail spermatophore. Actual length: 0.6 mm (from H. Mayer, 1957).

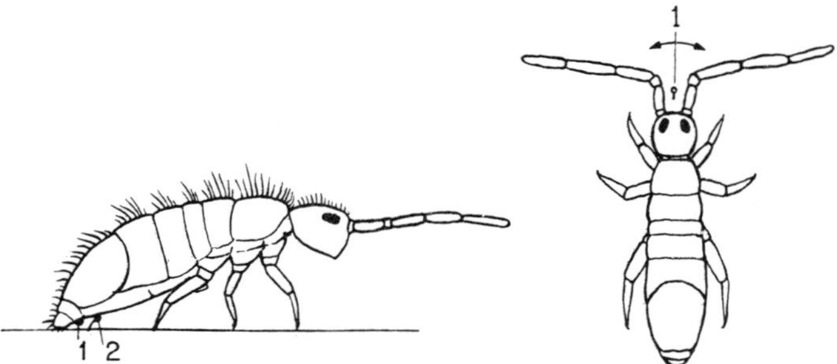

FIG. 88. "Grazing" springtail female of the genus *Orchesella*. Droplet of fluid (1) exuded from sexual aperture helps the female to pick up the spermatophore (2).

FIG. 89. Male springtail (*Orchesella*) wagging his head (see arrows) while testing the age of a spermatophore (1).

biologist has placed it there—a male will ignore it after the first "feel." If it turns out to be another male, it is pushed aside with a mighty blow. But if the object turns out to be a female, the pursuer will stay by her side even though she fails to pay the slightest attention to him. He keeps touching and circling round her, while following her every movement. The moment she comes to rest, he starts "planting" stalked spermatophores all around her, until she finds herself inside a sort of picket fence. Though the fence is loose and far from even, she is more likely than not to find her path blocked by one of the spermatophores. If her ova are mature, she will invariably pick one up.

This instance of indirect spermatophore transfer is highly interesting because it may be considered the one from which all the others have evolved. In species in which the male lacks a true copulative organ, there can clearly be no mating in the normal sense of the word; instead, there is some sort of "understanding" or "half-pairing." This is most obvious in the *Dicyrtomina* springtail, the female of which obviously lacks all the attitudes and reactions one might legitimately expect of a sexual partner.

Both *Sminthurides* and *Dicyrtomina* springtails have globular bodies in which the individual segments can barely be distinguished, another clear sign that they must be secondary forms. The "normal" springtails have an elongated, clearly segmented body. The two groups are known as the Symphypleona and the Arthropleona, respectively. While all the Symphypleona have a long "jumping fork" (furca), the length of this abdominal appendage varies a great deal in the Arthropleona. On Fig. 14c we saw that many soil-inhabiting species have lost their forked processes completely.

More About the Mating of Springtails

The sexual behavior of the Arthropleona resembles that of beetle mites. The males deposit stalked spermatophores even in the absence of females—a single male can produce considerably more than 100 in 2 or 3 days. The stalks are made of a secretion that hardens the moment it comes in contact with the air. In this respect, too, springtails and beetle mites resemble each other and contrast sharply with the scorpions, which, as the reader will remember, produce their spermatophores in special "molds." Springtail spermatophore stalks differ from those of beetle mites in being straight rather than wavy and in lacking a terminal beaker. The spermatophores cling to the stalks by virtue of their great viscosity, much as a drop of dew will cling to the hairtips of plants. At given times the ground is completely covered with these structures, for springtails are inordinately numerous. H. Mayer came across one such spermatophore "lawn" in Mainz and thought it looked like a thick blanket of mildew.

In these "gardens of love," springtail females can be seen promenading about (Fig. 88). If their ova are mature, they keep lowering their abdomens and wiping them along the ground. Even though this does not enable them to bring their genital aperture into contact with a particular spermatophore, they are bound to pick one up as they move through the dense lawn. Thus, they pick up the sperm more or less at random. Their genital aperture has a lateral slit, which undoubtedly assists the collection of sperm (males have only a minute longitudinal slit).

The males, too, do not walk about the spermatophore garden in their usual manner; indeed, they behave more peculiarly than even the females. From time to time they deflect their antennae and wag their heads to and fro (Fig. 89). Between the basal parts of their antennae on such occasions we invariably discover a spermatophore which they have been "testing." The males eat certain spermatophores but not others. The moment they have devoured one, they produce another. By keeping a precise record H. Mayer was able to report that they never swallow spermatophores less than 8 hours old. After that time the spermatozoa apparently lose their power and have to be replaced. It will be recalled that much the same thing happened in *Polyxenus*. Beetle mites, too, eat their own spermatophores.

The Arthropleona, likewise, do not form pairs in the true sense of the word. The sexes hardly take any notice of each other and act quite independently. The males like to touch the females and to keep near them, but they plant their spermatophores even if they are completely alone.

Since living males and females cannot be easily distinguished, the study of the sexual behavior of these springtails has proved extremely difficult. From a personal investigation of this behavior I can comment on some of the practical aspects of this kind of research.

I kept my springtail specimens—males and females of the species *Orchesella villosa* (Fig. 90)—in isolation and under constant illumination—so as to accustom them

FIG. 90. A 5-mm long springtail (*Orchesella villosa*) with black and yellow markings. Spermatophores of the primitive insect order Collembola were first discovered in this litter-inhabiting springtail (from Lubbock, 1930).

to light. When every animal had been given a number, I put them together, two at a time. I then began to keep watch on them in the confident expectation that among the hundreds of combinations I had produced, I should find a pair of willing partners. But for many weeks my patience and perseverance remained unrewarded. I was almost ready to give up when I noticed that two of the animals, which had been together for some time, kept stopping in different places, deflecting their antennae and shaking their heads violently. It was more by chance than design that I looked at their halting places under a more powerful lens and spotted a shiny droplet on a tall stalk. What I had before me was not a pair, but two males busily testing and replacing their spermatophores. After a few hours of closer observation, the whole process became clear. I noticed a repeated dabbing action, by which the males attach their stalks to the ground before slowly lifting their abdomens again, remaining in that position for a short time, and then walking off. Wherever they had performed this rite there was now a droplet on a stalk. Thereupon I put in more animals and observed the presumed females among them "grazing" on the lawn of spermatophores. I was even able to predict the time when they would lay their eggs, for they invariably did so within 1 or 2 days of "grazing," but never without it. Before long I could distinguish males from females by their external appearance—the males being invariably larger and fatter.

Until recently, all that was known of indirect spermatophore transfer was based on the studies of the sexual behavior of pseudoscorpions made by the English biologist Kew and the French biologist Vachon. Their discoveries seemed to deal with exceptional animal behavior. But when Sturm published his work on springtails, and F. Pauly described the strikingly similar behavior of the beetle mite (*Belba*), it became probable that this kind of sexual behavior was the rule rather than the exception among soil-inhabiting arthropods.

Some Phylogenetic Considerations

Now that we know more about indirect spermatophore transfer, what was once surprising seems perfectly natural. It is easy to see that life in the soil facilitates this form of sperm transfer, inasmuch as the soil is a semiaquatic habitat, its air saturated with moisture. Fluid spermatozoa do not dry up in it as quickly as they do above ground, so that they can be left safely in the open—

at least for some time. That such roundabout methods are used at all is probably due to the history of the development (phylogenesis) of these soil dwellers. Biologists are certain that all forms of animal life originated in the water and that the ancestors of modern soil arthropods emerged from an aquatic environment. As far as the crustaceans and arachnids among them are concerned, this has been fully corroborated by paleontological findings, and there is no reason to doubt that the same applies to the myriopods and insects as well. As we know, the males of many aquatic animals fertilize the eggs outside the body of the female. To do so they need not necessarily produce spermatophores, for the spermatozoa of most aquatic species can swim about in the water until they meet a free-floating ovum and fertilize it. Here, too, there must be some sort of "understanding" between the sexual partners—the ova must not be released too far from the males, for the tiny spermatozoa cannot range more than relatively short distances. On land, such an "understanding" seems even more essential, for the sperms must be deposited in closed parcels which prevent their soaking away or withering and facilitate their discovery and collection by the female (Fig. 91). The best solution, of course, is direct sperm transfer from the male to the female, but this solution has obviously not yet been "found" by many soil animals.* Instead, they have developed behavior patterns that ensure the transfer of sperm within a given territory and period of time. Only beetle mites, springtails, and the Symphyla myriopods (*Scutigerella*) lack

*Of particular phylogenetic interest is the mating behavior of the stump-footed tropical arthropod *Peripatus* (Onychophora; see p. 73). Here the male simply attaches hundreds of spermatophores to the female's skin, which dissolves in contact with the sperms, thus enabling them to swim to the ovaries through the body cavities. Onychophora females, accordingly, give birth to live offspring.

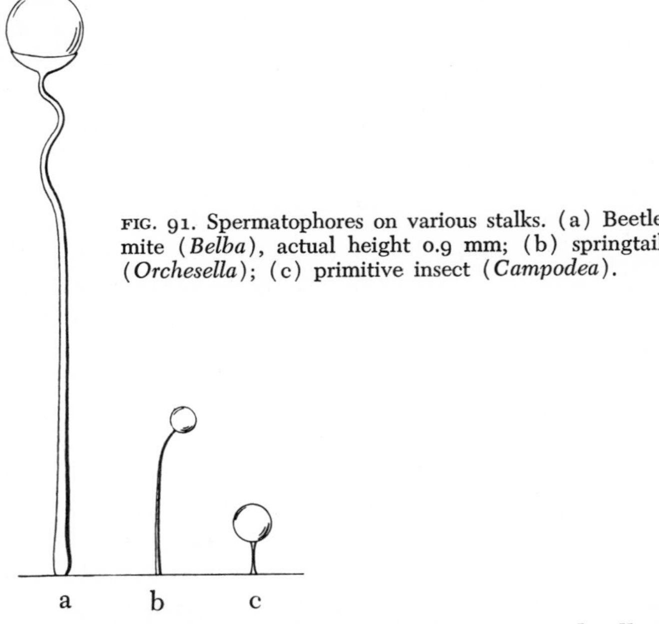

FIG. 91. Spermatophores on various stalks. (a) Beetle mite (*Belba*), actual height 0.9 mm; (b) springtail (*Orchesella*); (c) primitive insect (*Campodea*).

any kind of mutual "understanding." As true dwellers in the moist caverns of the earth they can behave as if they still lived in the water.

It is significant for their evolutionary history that Symphyla myriopods (see pp. 103-5) and springtails produce their spermatophores in such similar ways. Morphologists who study their anatomical structure and taxonomists who classify them are agreed that all insects are derived from myriopod ancestors. For many years their reasons for this assumption were purely anatomical and phylogenetic. More recently, behavioral studies have added support to their view, particularly when it became clear that yet another group of soil-dwelling primitive insects (the Campodeids, close relatives of the Japygids) also produce stalked spermatophores (Fig. 91). It must not be forgotten, however, that the springtails alone offer excellent confirmation of our phylogenetic hypothesis, providing us as they do with an evolutionary graduated scale of "mating" patterns:

1) Arthropleona (no pairing)
2) Symphypleona ("half"-pairing)
3) *Sminthurides aquaticus* (firmly attached pairs).

To the biologist the teeming host of animals beneath his feet is not merely one more proof of the inexhaustible variety of living phenomena, but also a challenge to put them into some kind of order, and to discover what is essential in, and typical of, them. Not only have we learned something about their diverse behavior patterns and their diverse structures, but we have also encountered a phenomenon common to most of them—indirect spermatophore transfer—and have seen what reasons of phylogeny and ecology—evolutionary history and interaction with the environment—explain its specific occurrence among soil animals.

But in other ways, also, soil animals constitute a varied yet fairly closed society—we have only to remember their special senses. Few other animal communities live under such uniform conditions and in so clearly limited a habitat. Not that the soil and its animal inhabitants constitute a world apart, for, as we saw in the chapter on the biosynthetic importance of soil animals, the soil itself depends on outside supplies for the maintenance of its metabolic balance. The animals, in particular, depend on the constant addition of material from the surface. In a strict sense this "living community" is not self-contained, with a marked ability for self-regulation, but only part of such a community. This is best seen in cultivated fields, where man has removed the natural "suppliers" and interrupted the organic line of supply. Both qualitatively and quantitatively the soil fauna has become greatly impoverished.

V
SOIL DESERTERS

Conversely, the soil animals maintain outgoing lines of communication with the outside world. This is particularly true of bark-inhabiting arthropods. We met one of these—a small myriopod (*Polyxenus lagurus*, p. 99). Other prominent bark dwellers are present among the springtails and mites. In tropical rain forests these animals have been responsible for the development of a new type of community in the epiphytic layers and humus deposits on the branches of trees (Fig. 92). Here are not only edaphic springtails, mites, and myriopods living more than 100 feet above the ground, but also worms and scorpions—all living side by side with true tree dwellers, which differ radically in appearance from "normal" soil animals (Fig. 93).

Another soil animal that has crossed its normal boundary is the springtail *Sminthurides aquaticus.* As we saw, it lives on the surface of the water. Its habitat is shared by another springtail *(Podura aquatica)* and by various spiders that are adept at running across the water. Some animals that were originally inhabitants of the soil have even moved *into* the water, among them numerous mites, and, especially, the aquatic oribatids, which have retained a striking number of their original characteristics. They look like ordinary beetle mites, except that their pseudostigmatic organs are much reduced (see p. 58). The so-called water mites have become truly aquatic animals, and either crawl about

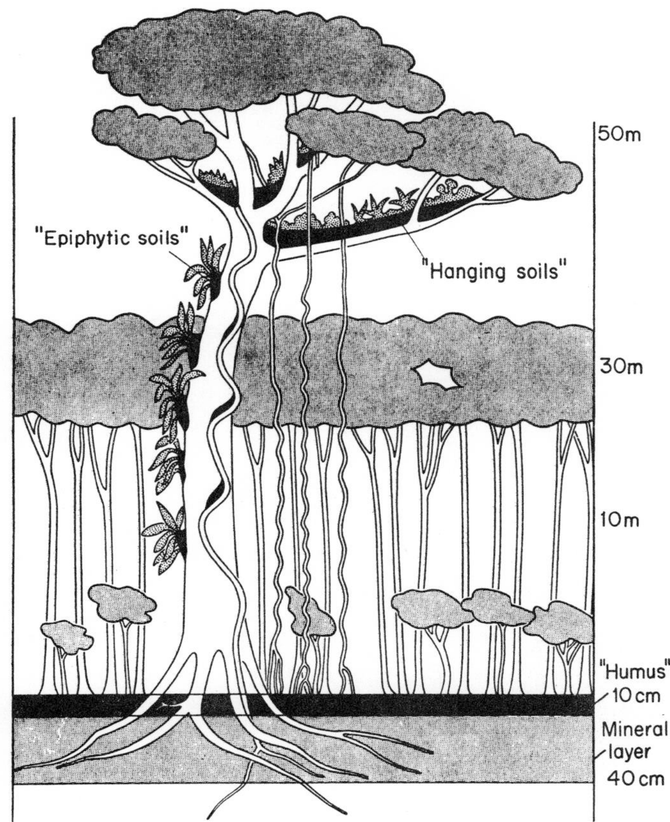

FIG. 92. The "hanging" soils of tropical jungles (from Delamore-Deboutteville, 1951). The black deposits on the lianas and branches consist of humus and contain a host of soil animals.

(Halacaridae) in water-laden soil or else swim about (Hydrachnellae) freely (Figs. 94-96). The Halacaridae transfer their spermatophores indirectly. Thus, the males of *Arrenurus* attach themselves to their partners by means of a sticky secretion and deposit a stalked spermatophore, which the female collects after appropriate rubbing motions by her partner. As K. Böttner has shown, Hydrachnellae, too, have a finely graduated scale of mating

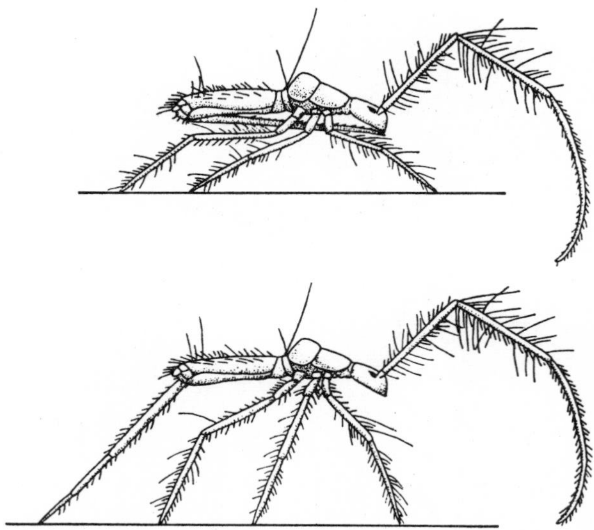

FIG. 93. The tree-dwelling springtail (*Campylothorax*), with extended furca (below).

patterns, and in these they resemble the springtails we have been discussing.

An unusual escape path from the soil, finally, leads straight into the snow and the ice. This is the path taken particularly by snow insects or glacier animals. The insects include various species of the genus *Isotoma* (Collembola), which regularly crawl about in the snow. Even more typical is the winter fly (*Boreus hiemalis*), a black wingless insect with spidery legs, which belongs to the scorpion flies (Fig. 97). Adults live only in the winter, when they are often seen in the snow. The extreme example, however, is the glacier "flea" (*Isotoma saltans*) which is actually a springtail (Fig. 98). This animal is so well adapted to life in ice and snow that it cannot survive away from glacial zones. Its food consists chiefly of wind-borne pollen grains. It only leaves the ice when it deposits its eggs—on stones. Its favorite temperature is below 3° C, and the animal carefully

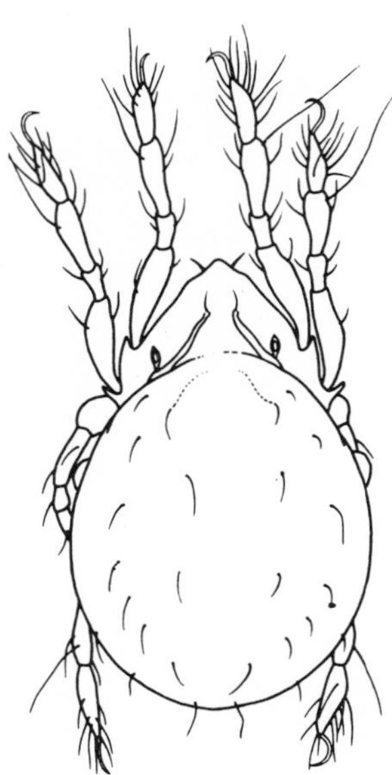

FIG. 94. Aquatic beetle mite (*Hydrozetes*) (from L. Beck, unpublished).

avoids temperatures above 12° C. Thus, it, too, keeps to constant temperature conditions, in this way resembling its relatives in the soil, and these, as we know, live under relatively cool conditions. Their thermal minimum, however, has become the glacier flea's thermal optimum. Glacier fleas hibernate no more than do other soil arthropods, for neither in the ice nor in the soil does the temperature ever drop as sharply as it does in the air —even during the severest frosts. In winter many soil animals actively move about under the snow, and glacier fleas have been found alive in the ice under yards of snow. Nor do most of the soil creatures come to harm when they are surprised by an occasional frost. Only earthworms are the exception. These animals, which die at temperatures of −1.5° to −2° C, regularly spend the winter in the deeper, and warmer, layers of the soil.

While most soil animals are tolerant of cold, they are far less adapted to life in an extremely dry and arid environment, such as in sand dunes or deserts. In southern

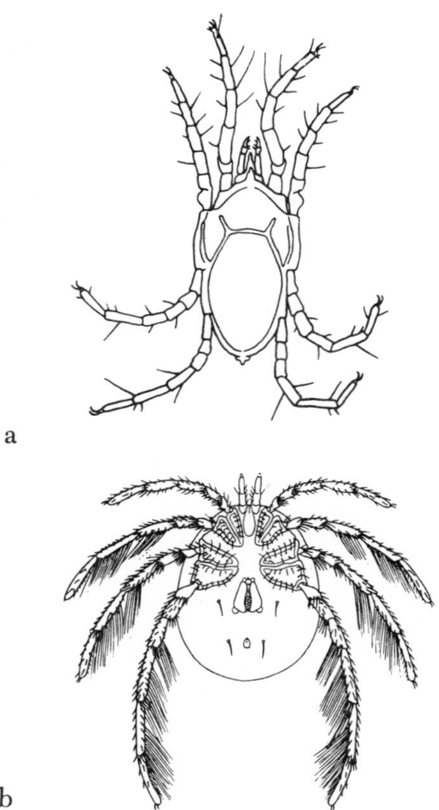

FIG. 95. (a) A mite (*Lobohalacarus*) inhabiting the bottom of rivers in the Harz mountains (from S. Husmann, 1959). (b) Water mite.

regions, however, a family of black beetles (Tenebrionidae) has a special predilection for this type of habitat. These beetles spend the day in the deeper layers of the sand and come out in search of food at night. Other sand lovers are the more northerly soil bug (*Brachypelta aterrima*), which we met in connection with its symbiotic nursing activity (cf. p. 82), some of its relatives (Cydnidae), and two carabid beetles (*Broscus cephalotes* and *Harpalus rufus*) which at night run about in, for

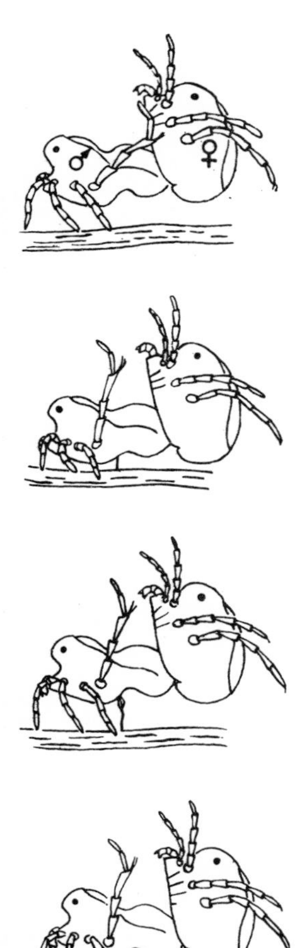

FIG. 96. Mating of a water mite (*Arrenurus globator*). The male approaches the female from below and attaches himself to her body with a special secretion. He then produces a stalked spermatophore and, by rubbing movements, ensures its reception by the female.

instance, flooded sand dunes where animals are never seen during the day. Their nocturnal habit becomes understandable when we realize that the daily temperature of the exposed sand surface may rise to more than 60° C.

Moist sea sand harbors completely different types of animals. Marine and terrestrial forms live side by side on the shore in close proximity. The best known of these are several amphipods (*Talitrus* and *Orchestia*), two

FIG. 97. The winter fly (*Boreus hiemalis*), which belongs to the order of scorpion flies (from H. Strübing, 1958). Actual size *ca.* 1 cm.

sandhoppers that are abundant in seaweed and other tide wrack on the shore, and rove beetles of the genus *Bledius* (Staphylinidae). The amphipods are true water creatures, but the beetles are true soil dwellers. Together with many other species, they settle the seashore from both sides as "soil animals" of a special kind, living on organic debris and on algae which flourish in the moist sand. Rove beetles build vertical tubes in the sand, on the walls of which they graze. They also have a remarkable way of nursing their eggs (Fig. 99).

The depth to which soil animals can inhabit the ground is another "border line" problem. In the southern Urals of the U.S.S.R. earthworms (*Allolobophora mariupolensis*) drive burrows down to 25 feet in the hard clay. Soil-dwelling termites and ants can also drive passages several yards down. In the arid tropics the galleries of termites invariably communicate with tunnels running down to the water table, which may be as much as 17 feet down. In general, however, soil animals—particularly the smaller ones—inhabit a relatively narrow zone below the surface.

A very characteristic exception is the fauna of cemeteries, which has an additional population layer at a depth of about 5 feet. This layer is mainly inhabited by springtails *(Onychiurus)*.

Finally, there are two channels by which soil inhabitants can leave their habitat and by which "strangers" can enter it in their turn—transport by clinging to other animals (phoresy) and parasitism. As mentioned, the

a

b

FIG. 98. The glacier flea (*Isotoma saltans*). (a) Side view. (b) Top view. Actual size 1.4 mm.

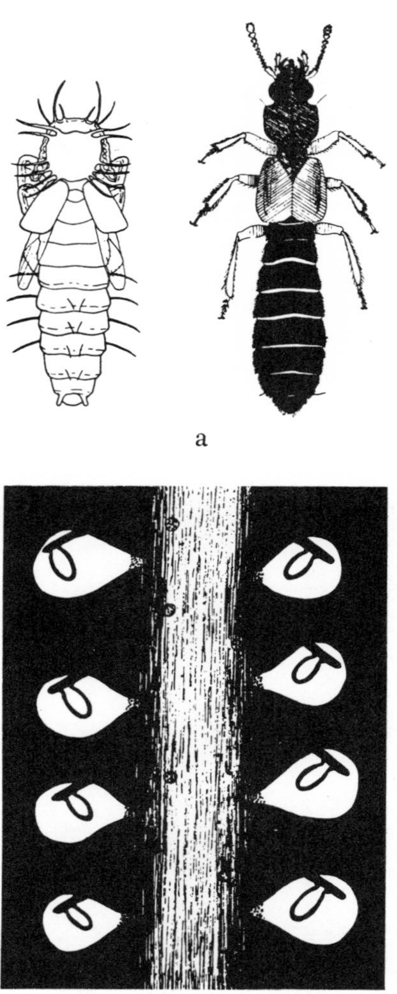

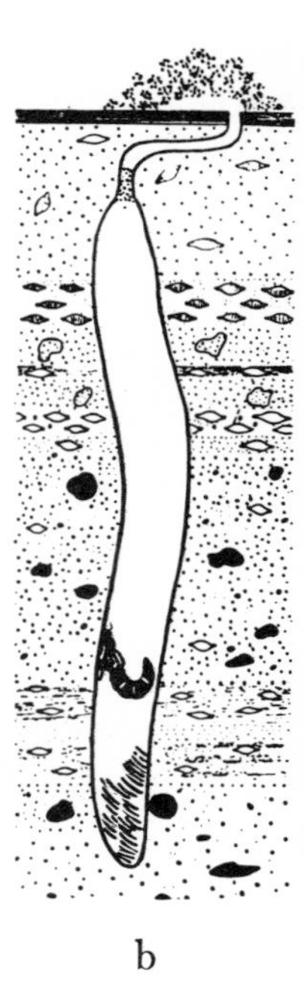

FIG. 99. (a) The beach-dwelling rove beetle (*Bledius spectabilis*) (from Wohlenberg, 1937). (a) Right: adult; left: pupa. (b) The tunnel of the rove beetle. The upper end of the tunnel cuts across the cover of blue algae (Cyanophyceae) at the surface of marshy coasts. (c) The "egg chambers" to the right and left of the tunnel. The ova are suspended from special stalks which afford them protection from moisture and fungal attack.

beetle mite (*Parasitus coleoptratorum*) travels as a "stowaway" on dung beetles, which convey it from one dung heap to the next (Fig. 100). This mode of travel enables soil-bound animals to break the bounds of their narrow habitat for a time. Numerous dung- and refuse-inhabiting nematode worms also cling to dung beetles as a means of transport. Other mites and nematodes are not content with using the beetles merely as transportation, but also feed on them—they become parasites. Parasitism is particularly prevalent in nematodes, above all among the land hair-worms (mermitids), which spend their larval stages in insects, but migrate into the soil on attaining sexual maturity—or rather remigrate, since they originally had been free-living "terrestrial" nematodes.

FIG. 100. The dung beetle (*Geotrupes*), covered with beetle mites (*Parasitus coleoptratorum*). The mites use the beetle as a means of transport between dung heaps (from A. Rapp, 1959).

VI
CONCLUSION

Even so apparently closed a habitat as the soil has a great many open ecological and biological frontiers. The soil animals are far from forming a closed community—we have only to remember that many insects spend only their larval or pupal stages below ground. Since the larval stage, however, constitutes the most important and longest phase in the life of such animals as cockchafers, noctuid moths, or cicadas, we may well include these in the soil community. The cockchafer grub spends 4 years in the ground, whereas the adult lives at most 4 weeks in a tree; the contrast is even more marked in certain cicadas—one of these, the North American seventeen-year "locust" (*Tibicen septendecim*), spends 17 years in the soil and only a few weeks above ground.

Our review of soil animals could be extended in several directions. In the living world there are no sharp boundaries. Soil animals form no exception to the general rule and are no more than an artificial division of the animal kingdom. Nevertheless, they comprise a vast spectrum of organic life and have helped us to touch on all the great problems of biology: phylogeny, structural diversity, growth and reproduction, metabolism, sensory perception and behavior, environment, individual and community life. This world of tiny hidden creatures proves to be a microcosm reflecting the forms and behavior patterns in the animal world at large. The biologist is not surprised that it should be so; he is used to discovering the laws of life in all sorts of inconspicuous and modest examples.

This little book will have served its purpose if it helps the layman to gain a better understanding of the wondrous world concealed beneath his feet.

ACKNOWLEDGMENTS

The illustrations listed below are from the following works:

Figs. 2, 4, 19, 20, and 21—Balogh: *Lebensgemeinschaften der Landtiere*. Berlin, 1958

Fig. 5a—Kükenthal: *Handbuch d. Biol.*, Vol. II, Part 1, 1933

Figs. 5b and 44—Trappmann: Thesis, Brunswick Technical High School, 1954

Figs. 6, 9c, 13, 15, and 26a—Kühnelt: *Bodenbiologie*. Vienna, 1950

Fig. 9a—*Bull. Biogeograph. Soc. Japan*, 1959

Figs. 11a, 18, 31b—A. Brauns: *Terricole Dipterenlarven*. Göttingen, 1954

Fig. 17—Zachariae: Lecture notes. Brunswick, Dec. 7, 1959

Fig. 22—Tischler: *Synökologie der Landtiere*. Stuttgart, 1955

Figs. 26b, 92, and 93—Delamare-Deboutteville: *Microfaune du sol*. Paris, 1951

Figs. 27 and 85—Stach: *The Apterygotan Fauna of Poland*. Cracow, 1954-56

Fig. 28—Kükenthal: *Handbuch d. Zool.*, Vol. II, Part 2, 1934

Fig. 35—Hesse-Doflein: *Tierbau und Tierleben*, Vol. II. Jena, 1943

Fig. 36b—*Zoologische Jahrb.* (Syst. Div.), Vol. 82, 1953

Fig. 37—*Zoologische Jahrb.* (Syst. Div.), Vol. 88, 1961

Figs. 38 and 53—*Zoologische Jahrb.* (Syst. Div.) Vol. 86, 1958

Figs. 42c and 42d—*Kosmos*, Vol. 56, 1960

Fig. 43—*Brehms Tierleben*

Figs. 45, 62, 63, and 64—*Zool. Jahrb.* (Syst. Div.), Vol. 84, 1956

Figs. 9d, 46, and 100—*Zool. Jahrb.* (Syst. Div.), Vol. 86, 1959

Figs. 47 and 51b—Meisenheimer: *Geschlecht und Geschlechter im Tierreich*, Vol. I. Jena, 1921

Figs. 49, 74, and 75—*Zeitschr. f. Tierpsychol.*, Vol. 17, 1960

Fig. 50—*Zeitschr. f. Morphologie und Ökologie der Tierre*, Vol. 45, 1957

Figs. 54, 55, 56, and 57—*Zeitschr. f. Tierpsychol.* Vol. 14, 1957

Figs. 58, 59, and 60—*Naturwissenschaften*, Vol. 45, 1958

Fig. 61—Von Buddenbrock: *Das Liebesleben der Tiere*. Bonn, 1953

Figs. 65, 66, 67, and 68—*Zool. Jahrb.* (Syst. Div.), Vol. 84, 1956

Figs. 69 and 70—*Comptes rendus, Acad. des Sci.*, Vol. 249, 1959

Figs. 71, 72, and 73—*Verh. d. deutsche Zool. Gesellschaft*, 1959

Figs. 77, 78, 79, and 80—*Zeitschr. f. Tierpsychol.*, Vol. 12, 1955

Figs. 81, 82, 83, and 84—*Zeitschr. f. Tierpsychol.*, Vol. 13, 1956

Fig. 86—Schulze: *Biologie d. Tiere Deutschlands*, Part 25, 1926

Fig. 87—*Zool. Jahrb.* (Syst. Div.), Vol. 85, 1957

Fig. 90—Kükenthal: *Handbuch d. Zool.*, Vol. IV, 1930

Fig. 95a—*Gewässer und Abwässer*. Düsseldorf, 1959

Fig. 96—Kükenthal: *Handbuch d. Zool.*, Vol. III, 1941

Fig. 97—Strübing: *Schneeinsekten*, 1958

99—*Helgoländer wissenschaftliche Meeresuntersuchungen*, Vol.

INDEX